Metals and the Biology and Virulence of Brucella

R. Martin Roop II · Clayton C. Caswell
Editors

Metals and the Biology
and Virulence of *Brucella*

 Springer

Editors
R. Martin Roop II
Department of Microbiology and
  Immunology, Brody School of Medicine
East Carolina University
Greenville, NC
USA

Clayton C. Caswell
Department of Biomedical Sciences and
  Pathobiology, Virginia–Maryland College
  of Veterinary Medicine
Virginia Tech
Blacksburg, VA
USA

ISBN 978-3-319-53621-7   ISBN 978-3-319-53622-4   (eBook)
DOI 10.1007/978-3-319-53622-4

Library of Congress Control Number: 2017937114

Printed on acid-free paper

This Springer imprint is published by Springer Nature
The registered company is Springer International Publishing AG
The registered company address is: Gewerbestrasse 11, 6330 Cham, Switzerland

Contents

1 **Introduction and Overview** 1
R. Martin Roop II and Clayton C. Caswell
  1.1 *Brucella* 1
    1.1.1 *B. melitensis*, *B. abortus* and *B. suis* 2
    1.1.2 *B. canis* and *B. ovis* 3
    1.1.3 Other *Brucella* Species 3
  1.2 Biological Functions of Metals and the Importance of Metal
       Homeostasis in Living Cells 4
  1.3 Metal Homeostasis in *Brucella* Strains 5
  References 6

2 **Iron** 9
R. Martin Roop II, Ahmed E. Elhassanny, Marta A. Almirón,
Eric S. Anderson and Xavier J. Atkinson
  2.1 Fe as a Micronutrient for *Brucella* Strains 9
  2.2 The Mammalian Host Is an Fe-Deprived Environment 10
  2.3 Fe Acquisition by *Brucella* 13
    2.3.1 Siderophore Production 13
    2.3.2 Fe^{3+}-Siderophore Transport Systems in *Brucella* 17
    2.3.3 Siderophore Production and Virulence of *Brucella*
           abortus in Pregnant Ruminants 19
    2.3.4 Heme as an Fe Source for *Brucella* Strains 20
    2.3.5 Ferrous Iron Transport by *Brucella* Strains 22
    2.3.6 Siderophore-Independent Transport of Fe^{3+}
           in *Brucella* 25
  2.4 Detoxification of Excess Intracellular Fe by *Brucella* Strains 25
    2.4.1 Bacterioferritin and Dps 25
    2.4.2 MbfA 27

2.5   Regulation of Fe Homeostasis in *Brucella*   28
      2.5.1   Irr   28
      2.5.2   RirA   29
      2.5.3   DhbR and Other Fe Source-Specific Regulators   31
      2.5.4   BsrH   32
2.6   Conclusions   33
References   33

3   Manganese   41
      R. Martin Roop II, Joshua E. Pitzer, John E. Baumgartner
      and Daniel W. Martin
3.1   Manganese as a Micronutrient and Bacterial Mn Transport   41
3.2   MntH-Mediated Mn Transport Is Critical for the Basic Biology
      and Virulence of *Brucella* Strains   42
3.3   Mn-Dependent Proteins that Contribute to the Virulence
      of *Brucella* Strains   45
      3.3.1   SodA   45
      3.3.2   PykM   47
      3.3.3   Rsh   49
      3.3.4   Pcs   50
      3.3.5   BpdA and BpdB   50
3.4   Manganese as an Antioxidant   51
3.5   Manganese Homeostasis   51
      3.5.1   Mur   53
      3.5.2   EmfA   54
3.6   Conclusions   54
References   54

**4   The Role of Zinc in the Biology and Virulence
      of *Brucella* Strains**   63
      Clayton C. Caswell
4.1   Zinc Import by ZnuABC   64
4.2   Zinc Export by ZntA   66
4.3   Zinc-Responsive Transcriptional Regulators Zur and ZntR   66
4.4   Zinc-Dependent Proteins in *Brucella*   68
      4.4.1   The Cu-Zn Superoxide Dismutase C (SodC)   68
      4.4.2   The Transcriptional Regulatory Protein MucR   69
      4.4.3   The Type IV Secretion System Effector Protein RicA   69
      4.4.4   Other Potential Zinc-Containing Proteins in *Brucella*   70
References   71

5   Nickel Homeostasis in *Brucella* spp.   73
      James A. Budnick and Clayton C. Caswell
5.1   Nickel Import by NikABCDE and NikKMLQO   73
5.2   The Nickel and Cobalt Exporter RcnA   76

5.3 Nickel-Responsive Regulators NikR and RcnR 76
5.4 Nickel-Dependent Proteins in *Brucella* 78
      5.4.1 Urease 78
      5.4.2 Other Potential Nickel-Containing Proteins
             in *Brucella* 79
References 79

6 **Magnesium, Copper and Cobalt** 81
R. Martin Roop II, John E. Baumgartner, Joshua E. Pitzer
and Daniel W. Martin
6.1 Magnesium 81
      6.1.1 CorA, MgtE and MgtB 82
      6.1.2 MgtC 83
6.2 Copper 84
6.3 Cobalt 86
6.4 Cobalamin Transport 89
6.5 Conclusions 90
References 90

Chapter 1
Introduction and Overview

R. Martin Roop II and Clayton C. Caswell

Abstract Bacteria in the genus *Brucella* are important human and veterinary pathogens, and they require a variety of metals to support their physiology and metabolism. Mammals employ both metal limitation and metal intoxication as defenses against invading pathogens, and correspondingly the metal acquisition and detoxification systems of *Brucella* strains play essential roles in their virulence.

Keywords *Brucella* · Zoonosis · Metals · Metal toxicity

1.1 *Brucella*

The genus *Brucella* currently consists of 11 recognized species of Gram-negative bacteria. Five of these—*B. melitensis*, *B. abortus*, *B. suis*, *B. canis* and *B. ovis*—are important veterinary and human pathogens. Molecular studies have shown that all *Brucella* strains are closely related at the genetic level. The separate 'species' designations have been retained, however, because these bacteria can be subdivided into distinct phenotypic groups (e.g., species and biovars within these species) that display different host specificities and virulence properties. These distinctions are important for understanding the epidemiology and pathogenesis of *Brucella* infections (reviewed in Whatmore 2009).

R.M. Roop II (✉)
Department of Microbiology and Immunology, Brody School of Medicine,
East Carolina University, Greenville, NC 27834, USA
e-mail: roopr@ecu.edu

C.C. Caswell
Department of Biomedical Sciences and Pathobiology,
Virginia–Maryland College of Veterinary Medicine, Virginia Tech,
Blacksburg, VA, USA
e-mail: caswellc@vt.edu

© Springer International Publishing AG 2017
R. Martin Roop II and Clayton C. Caswell (eds.), *Metals and the Biology and Virulence of Brucella*, DOI 10.1007/978-3-319-53622-4_1

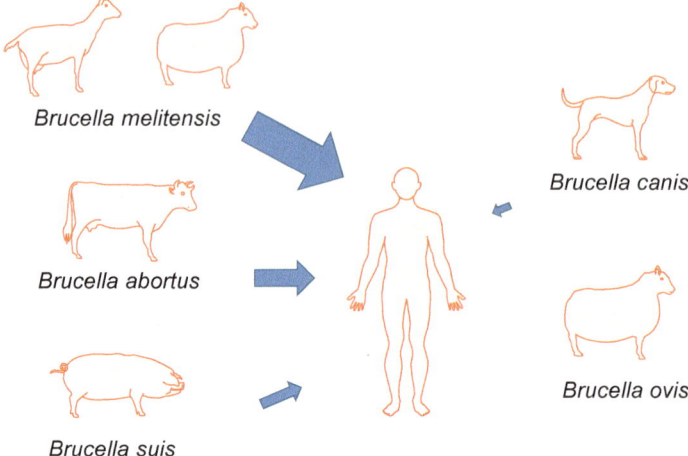

Fig. 1.1 Natural hosts of *Brucella melitensis*, *B. abortus*, *B. suis*, *B. canis* and *B. ovis* and zoonotic potential of these strains. The size of the arrow pointing from the natural hosts to humans indicates the relative propensity of each *Brucella* spp. to cause human disease

1.1.1   *B. melitensis, B. abortus and B. suis*

Brucella melitensis, *B. abortus* and *B. suis* cause abortion and infertility in goats and sheep, cattle, and swine, respectively (Atluri et al. 2011). These bacteria are highly infectious in their natural hosts where they can produce chronic, often life-long infections. Agricultural communities worldwide devote tremendous resources annually to prevent and control food animal brucellosis (Godfroid et al. 2014). *B. melitensis*, *B. abortus* and *B. suis* can also be readily transmitted to humans via the consumption of unpasteurized dairy products or direct contact with infected animals, where they produce a serious, chronic debilitating febrile disease (Fig. 1.1). Human brucellosis represents a major public health problem in areas of the world where the disease is not effectively controlled in food animals, and in fact, this disease is considered to be one of the world's leading zoonotic infections (Pappas et al. 2006).

B. melitensis, *B. suis* and *B. abortus* strains also have characteristics that make them attractive as agents of biowarfare or bioterrorism (Valderas and Roop 2006). Specifically, they have low infectious doses via the aerosol route, the disease they produce in humans is difficult to treat with antibiotics, and there is no vaccine that can be safely and effectively used to prevent human brucellosis. Historically, *B. melitensis* and *B. suis* strains were included in the bioweapons arsenals of several countries before the global movement to ban the use of these weapons in the late 1960s and early 1970s. Today, many countries still tightly regulate the possession of *B. melitensis*, *B. suis* and *B. abortus* strains due to their potential use in bioterrorism.

1.1.2 B. canis and B. ovis

B. canis causes abortion and infertility in dogs, and canine brucellosis is a serious concern in kennels (Wanke 2004). This bacterium can also be transmitted from dogs to humans as a zoonotic agent (Fig. 1.1). Although the reported incidence of human disease caused by *B. canis* is low compared to that caused by *B. melitensis*, *B. abortus* and *B. suis*, it has been proposed that many human *B. canis* infections likely go unrecognized (Krueger et al. 2014). *B. ovis* causes epididymitis is sheep and is an important cause of infertility in rams worldwide (Gouletsou and Fthenakis 2015), but these strains are not known to cause human disease.

1.1.3 Other Brucella Species

Limited information is available regarding the importance of the remaining *Brucella* species as pathogens. *B. microti*, for instance, produces severe and sometimes fatal disease in both wild rodents (Hubálek et al. 2007) and experimentally infected mice (Jiménez de Bagüés et al. 2010), and was recently isolated from a wild boar (Rónai et al. 2015). But it is unclear how common and widespread *B. microti* infections are in wild rodents (Hammerl et al. 2015) or other wildlife, and disease in humans or domestic animals associated with this strain has not been reported. Similarly, although *B. ceti* and *B. pinnipedialis* strains are routinely isolated from marine mammals (Foster et al. 2007), overt clinical signs appear to be uncommon in the animals from which these strains are isolated (Nymo et al. 2011; Guzmán-Verri et al. 2012). *B. ceti* strains, have however, been isolated from a

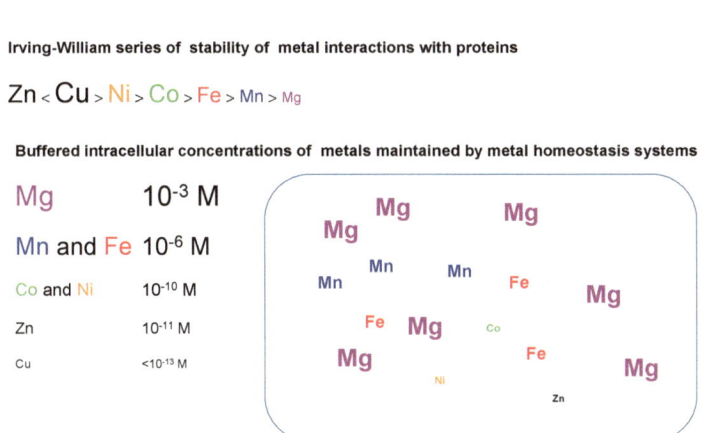

Fig. 1.2 Inverse correlation of the Irving-Williams series of stability of metal interactions with proteins (Irving and Williams 1948) and the buffered set points of intracellular metal concentrations maintained by metal homeostasis systems (Foster et al. 2014)

limited number of human infections (reviewed in Whatmore et al. 2008), suggesting that these strains have the potential to be zoonotic pathogens. Information regarding the prevalence, natural host range, pathogenicity and zoonotic potential of *B. neotomae* (Stoenner and Lackman 1957), *B. inopinata* (Scholz et al. 2010), *B. papionis* (Whatmore et al. 2014) and *B. vulpis* (Scholz et al. 2016) is even more limited, since only a few isolates of these strains have been characterized.

1.2   Biological Functions of Metals and the Importance of Metal Homeostasis in Living Cells

Copper (Cu), zinc (Zn), iron (Fe), manganese (Mn), magnesium (Mg), nickel (Ni) and cobalt (Co) serve as important micronutrients for living cells. It has been estimated that approximately 50% of all enzymes, and 1/4–1/3 of proteins in general, require metal co-factors for their activity (Waldron et al. 2009). Metals can play either catalytic or structural roles in protein function. Zn in the active site of the enzyme carbonic anhydrase, for instance, directly participates in the interconversion of CO_2 and HCO_3^-, a reaction that maintains cytoplasmic pH balance in both prokaryotic and eukaryotic cells (Supuran and Scozzafava 2007). The interaction of specific amino acid residues with Zn also coordinates the proper folding of the large family of eukaryotic proteins known as 'Zn finger' proteins (Klug 2010) which perform a variety of different biological functions. The redox activities of Fe and Cu make proteins containing these metals essential components of electron transport chains (Liu et al. 2014), and the Fe incorporated in heme plays a critical role in O_2 transport in mammals (Fujiwara and Harigae 2015).

Despite the fact that they are essential micronutrients, metals can also be toxic when their levels exceed those required to meet the physiologic needs of the cell (Summers 2009). To prevent metal toxicity, both prokaryotic and eukaryotic organisms have evolved finely-tuned homeostasis systems that tightly control the intracellular levels of metals (Waldron and Robinson 2009; Foster et al. 2014). These systems are typically comprised of metal importers and exporters, metal chaperones, and metal storage and detoxification proteins, and the expression of the genes that encode these proteins is tightly regulated in response to cellular metal levels. Metal toxicity occurs primarily for two reasons. First, the affinity of metals for proteins follows a scale known as the Irving-Williams series, with Cu and Zn having the highest affinity and Mg having the lowest (Waldron and Robinson 2009), and the intracellular levels of the high affinity metals must be maintained at lower levels that those of the lower affinity metals to avoid the higher affinity metals displacing the lower affinity metals in proteins or enzymes where the latter are essential for protein function, or dedicated metallochaperones must be in place to ensure that proteins acquire the proper metal. As shown in Fig. 1.2, these metal homeostasis systems 'buffer' the intracellular levels of the individual metals to ensure that their concentrations are inversely proportionate to their potential toxicity

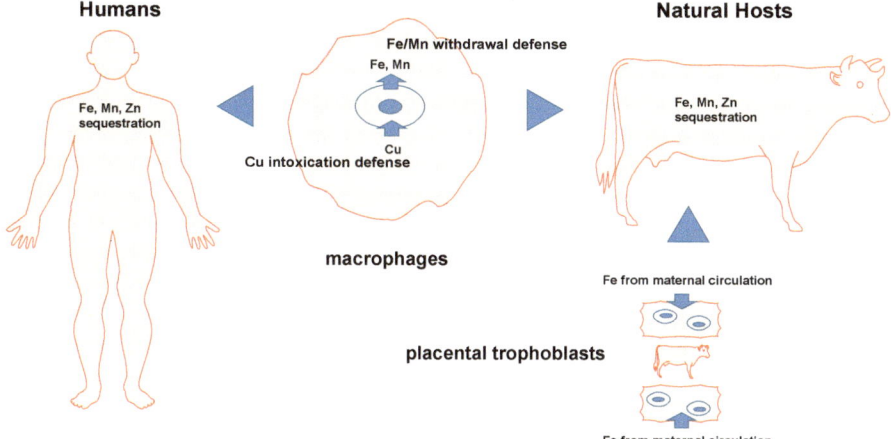

Fig. 1.3 Homeostasis systems and immune defenses that potentially influence the availability of metals to the brucellae during residence in the two key cell types they inhabit in their natural hosts and in humans

(Foster et al. 2014). The second reason for metal toxicity is that Fe reacts with reactive oxygen species such as the superoxide ion (O_2^-) to form highly reactive hydroxyl radicals which can damage proteins, nucleic acids and lipids (Imlay 2013). Consequently, Fe homeostasis systems also play important roles in oxidative defense, and the genes that encode the components of these homeostasis systems are often responsive to oxidative stress in addition to their regulation in response to cellular levels of the corresponding metal (Faulkner and Helmann 2011).

1.3   Metal Homeostasis in *Brucella* Strains

Early studies of the nutritional requirements of *Brucella* strains during in vitro cultivation determined that Fe and Mg were essential micronutrients (Gerhardt 1958). Because it is very difficult to remove contaminating metals from media components and culture vessels, however, these earlier studies underestimated the importance of other metals in the physiology of these bacteria. More recent studies employing genetically defined mutants and genome analysis have not only confirmed the importance of Fe and Mg as micronutrients for *Brucella* strains, but have also shown that Mn, Zn, Cu, Ni and Co play critical roles in their basic physiology (Roop et al. 2012).

Brucella strains live in close association with their mammalian hosts (Roop et al. 2009), where they reside predominantly as intracellular pathogens. Their capacity to survive and replicate in host macrophages underlies their ability to cause chronic infections, and their extensive intracellular replication in placental trophoblasts plays an important role in their capacity to induce abortion in their natural hosts. Mammals not only tightly control the levels of metals in their tissues to avoid toxicity, but they also employ Fe, Mn and Zn deprivation and Cu intoxication as mechanisms to limit the replication of microbial pathogens (Hood and Skaar 2012) (Fig. 1.3). It is also notable in this regard that one of the important roles that placental trophoblasts play during pregnancy is to provide Fe from the maternal circulation to the developing fetus (Carter 2012; de Oliveira et al. 2012). Thus, it is not surprising that metal homeostasis systems have been shown to play critical roles in the virulence of *Brucella* strains in both experimental and natural hosts (Roop 2012). The following chapters will review the information that is currently available regarding the role that metal homeostasis plays in the biology and virulence of *Brucella* strains.

References

Atluri VL, Xavier MN, de Jong MF, den Hartigh AB, Tsolis RM (2011) Interactions of the human pathogenic *Brucella* species with their hosts. Ann Rev Microbiol 65:523–541

Carter AM (2012) Evolution of placental function in mammals: the molecular basis of gas and nutrient transfer, hormone secretion, and immune responses. Physiol Rev 92:1543–1576

De BK, Stauffer L, Koylass MS, Sharp SE, Gee JE, Helsel LO, Steigerwalt AG, Vega R, Clark TA, Daneshvar MI, Wilkins PP, Whatmore AM (2008) Novel *Brucella* strain (B01) associated with a prosthetic breast implant infection. J Clin Microbiol 46:43–49

de Oliveira CM, Rodrigues MN, Miglino MA (2012) Iron transportation across the placenta. An Acad Bras Cienc 84:1115–1120

Faulkner MJ, Helmann JD (2011) Peroxide stress elicits adaptive changes in bacterial metal ion homeostasis. Antioxid Redox Signal 15:175–189

Foster G, Osterman BS, Godfroid J, Jacques I, Cloeckaert A (2007) *Brucella ceti* sp nov. and *Brucella pinnipedialis* sp. nov. for *Brucella* strains with cetaceans and seals as their preferred hosts. Int J Syst Evol Microbiol 57:2688–2693

Foster AW, Osman D, Robinson NJ (2014) Metal preferences and metallation. J Biol Chem Chem 289:28095–28103

Fujiwara T, Harigae H (2015) Biology of heme in mammalian erythroid cells and related disorders. BioMed Res Int 2015:e278536

Gerhardt P (1958) The nutrition of brucellae. Microbiol Rev 22:81–98

Godfroid J, De Bolle X, Roop RM II, O'Callaghan D, Tsolis RM, Baldwin C, Santos RL, McGiven J, Olsen S, Nymo IH, Larsen A, Al Dahouk S, Letesson JJ (2014) The quest for a true One Health perspective of brucellosis. Rev Sci Tech 33:521–538

Gouletsou PG, Fthenakis GC (2015) Microbial diseases of the genital tract of rams or bucks. Vet Microbiol 181:130–135

Guzmán-Verri C, González-Barrientos R, Hernández-Mora G, Morales JA, Barquero-Calvo E, Chavez-Olarte E, Moreno E (2012) *Brucella ceti* and brucellosis in cetaceans. Front Cell Infect Microbiol 2:e3

Hammerl JA, Ulrich RG, Imholt C, Scholz HC, Jacob J, Kratzmann N, Nöckler K, Al Dahouk S (2015) Molecular survey on brucellosis in rodents and shrews—natural reservoirs of novel *Brucella* species in Germany? Transbound Emerg Dis. doi:10.1111/tbed.12425

Hood MI, Skaar EP (2012) Nutritional immunity: transition metals at the pathogen-host interface. Nature Rev Microbiol 10:525–537

Hubálek Z, Scholz HC, Sedláček I, Meltzer F, Sanogo YO, Nesvadbová J (2007) Brucellosis of the common vole (*Microtus arvalis*). Vector-Borne Zoonot Dis 7:679–687

Imlay JA (2013) The molecular mechanisms and physiological consequences of oxidative stress: lessons from a model bacterium. Nature Rev Microbiol 11:443–454

Irving H, Williams RJ (1948) Order of stability of metal complexes. Nature 162:746–747

Jiménez de Bagüés MP, Ouahrani-Bettache S, Quintana JF, Mitjana O, Hanna N, Bessoles S, Sanchez F, Scholz HC, Lafont V, Köhler S, Occhialini A (2010) The new species *Brucella microti* replicates in macrophages and causes death in murine models of infection. J Infect Dis 202:3–10

Klug A (2010) The discovery of zinc fingers and their applications in gene regulation and genome manipulation. Annu Rev Biochem 79:213–231

Krueger WS, Lucero NE, Brower A, Heil GL, Gray GC (2014) Evidence for unapparent *Brucella canis* infections among adults with occupational exposure to dogs. Zoonoses Publ Health 61:509–518

Liu JS, Chakraborty S, Hosseinzadeh P, Yu Y, Tian S, Petrik I, Bhagi A, Lu Y (2014) Metalloproteins containing cytochrome, iron-sulfur, or copper redox centers. Chem Rev 114:4366–4469

Nymo IH, Tryland M, Godfroid J (2011) A review of *Brucella* infection in marine mammals, with special emphasis on *Brucella pinnipedialis* in the hooded seal (*Cystophora cristata*). Vet Res 42:e93

Pappas G, Papadimitriou P, Akritidis N, Christou L, Tsianos EV (2006) The new global map of human brucellosis. Lancet Infect Dis 6:91–99

Rónai Z, Kreizinger Z, Dán A, Drees K, Foster JT, Bányai K, Marton S, Szeredi L, Jánosi A, Gyuranecz M (2015) First isolation and characterization of *Brucella microti* from wild boar. BMC Vet Res 11:e47

Roop RM II (2012) Metal acquisition and virulence in *Brucella*. Anim Health Res Rev 13:10–20

Roop RM II, Gaines JM, Anderson ES, Caswell CC, Martin DW (2009) Survival of the fittest: how *Brucella* strains adapt to their intracellular niche in the host. Med Microbiol Immunol 198:221–238

Roop RM II, Anderson E, Ojeda J, Martinson D, Menscher E, Martin DW (2012) Metal acquisition by *Brucella* strains. In: López-Goñi I, O'Callaghan D (eds) *Brucella*—molecular microbiology and genomics. Caister Academic Press, Norfolk, UK, pp 179–199

Schlabritz-Loutsevitch NE, Whatmore AM, Quance CR, Koylass MS, Cummins LB, Dick EJ Jr, Snider CL, Cappelli D, Ebersole JL, Nathanielsz PW, Hubbard GB (2009) A novel *Brucella* isolate in association with two cases of stillbirth in non-human primates—first report. J Med Primatol 38:70–73

Scholz HC, Nöckler K, Göllner C, Bahn P, Vergnaud G, Tomaso H, Al Dahouk S, Kämpfer P, Cloeckaert A, Maquart M, Zygmunt MS, Whatmore AM, Pfeffer M, Huber B, Busse HJ, De BK (2010) *Brucella inopinata* sp. nov., isolated from a breast plant infection. Int J Syst Evol Microbiol 60:801–808

Scholz HC, Revilla-Fernández S, Al Dahouk S, Hammerl JA, Zygmunt MS, Cloeckaert A, Koylass M, Whatmore AM, Blom J, Vergnaud G, Witte A, Aistleitner K, Hofer E (2016) *Brucella vulpis* sp. nov., isolated from mandibular lymph nodes of red foxes (*Vulpes vulpes*). Int J Syst Evol Microbiol 66:2090–2098

Stoenner HG, Lackman DB (1957) A new species of *Brucella* isolated from the desert wood rat, *Neotoma lepida* Thomas. Am J Vet Res 18:947–951

Summers AO (2009) Damage control: regulating defenses against toxic metals and metalloids. Curr Opin Microbiol 12:138–144

Supuran CT, Scozzafava A (2007) Carbonic anhydrases as targets for medicinal chemistry. Biorg Med Chem 15:4336–4350

Valderas MW, Roop RM II (2006) *Brucella* and bioterrorism. In: Anderson B, Friedman H, Bendinelli M (eds) Microorganisms and bioterrorism. Springer, New York, NY, pp 139–153

Waldron KJ, Robinson NJ (2009) How do bacterial cells ensure that metalloproteins get the correct metal? Nature Rev Microbiol 6:25–35

Waldron KJ, Rutherford JC, Ford D, Robinson NJ (2009) Metalloproteins and metal sensing. Nature 460:823–830

Wanke MM (2004) Canine brucellosis. Anim Reprod Sci 82–83:195–207

Whatmore AM (2009) Current understanding of the genetic diversity of *Brucella*, an expanding genus of zoonotic pathogens. Infect Genet Evol 9:1168–1184

Whatmore AM, Dawson CE, Groussaud P, Koylass MS, King AC, Shankster SJ, Sohn AH, Probert WS, McDonald WL (2008) Marine mammal *Brucella* genotype associated with zoonotic infection. Emerg Infect Dis 14:517–518

Whatmore AM, Davison N, Cloeckaert A, Al Dahouk S, Zygmunt MS, Brew SD, Perrett LL, Koylass MS, Vergnaud G, Quance C, Scholz HC, Dick EJ Jr, Hubbard G, Schlabritz-Loutsevitch NE (2014) *Brucella papionis* sp. nov., isolated from baboons (*Papio* spp.). Int J Syst Evol Microbiol 64:4120–4128

Chapter 2
Iron

R. Martin Roop II, Ahmed E. Elhassanny, Marta A. Almirón, Eric S. Anderson and Xavier J. Atkinson

Abstract Fe is an essential micronutrient for *Brucella* strains. Meeting their physiologic need for this metal is especially challenging for these bacteria because they live in close association with their mammalian hosts, and Fe-sequestration is a well-documented host defense against microbial pathogens. The following chapter will describe what is presently known about Fe homeostasis in *Brucella* strains, and how the individual cellular components involved in this process contribute to virulence.

Keywords *Brucella* · Iron · Macrophage · Iron homeostasis

2.1 Fe as a Micronutrient for *Brucella* Strains

Iron (Fe) is a widely used prosthetic group in proteins due to its exceptional reactivity under physiologic conditions (Crichton 2009). Its capacity to undergo redox reactions, for instance, makes it an important component of the proteins that comprise electron transport chains as well as those that participate in maintaining the redox balance of cells. The cationic nature of Fe also allows this metal to stabilize substrates and reaction intermediates in the active sites of enzymes. Ferrous (Fe^{2+}) and ferric (Fe^{3+}) iron are the main oxidation states found in biological systems, and the redox potential of the Fe^{2+}/Fe^{3+} couple can vary greatly based on the ligands to

R.M. Roop II (✉) · A.E. Elhassanny
Department of Microbiology and Immunology, Brody School of Medicine, East Carolina University, Greenville, NC 27834, USA
e-mail: roopr@ecu.edu

M.A. Almirón
Instituto de Investigaciones Biotecnológicas–Instituto Tecnólogico de Chascomús, Universidad Nacional de San Martín–Consejo Nacional de Investigaciones Científicas y Técnicas, Buenos Aires, Argentina

E.S. Anderson · X.J. Atkinson
Department of Biology, East Carolina University, Greenville 27834, USA

© Springer International Publishing AG 2017
R. Martin Roop II and Clayton C. Caswell (eds.), *Metals and the Biology and Virulence of Brucella*, DOI 10.1007/978-3-319-53622-4_2

which this cation is bound (Kosman 2013). This makes Fe extremely versatile in biochemical reactions. Fe can be found in proteins as either the free metal or as a component of heme or Fe-S centers. Because of its versatility, Fe is an essential co-factor for many cellular enzymes, including those involved in DNA synthesis, central carbon metabolism, cellular respiration and signal transduction. With the notable exception of bacteria in the genus *Lactobacillus* (Archibald 1983) and *Borrelia* (Posey and Gherardini 2000), all living organisms that have been studied require Fe as a micronutrient. For instance, early studies showed that *Brucella* strains require at least 0.5 μM Fe for growth and grow best when supplied with Fe concentrations >1.8 μM (Waring et al. 1953), which is typical of Fe requirements that have been reported for other bacteria (Lengeler et al. 1999).

2.2   The Mammalian Host Is an Fe-Deprived Environment

Acquiring sufficient Fe to meet their physiologic needs is a particular challenge for *Brucella* strains because they reside predominately in close association with their mammalian hosts (Roop et al. 2009). One of the reasons for this is that the vast majority of the Fe in mammals is incorporated into proteins as heme, Fe-S clusters or mononuclear or dinuclear Fe centers, and is thus not readily accessible (Hood and Skaar 2012). Another consideration is that the soluble and biologically active form of Fe, Fe^{2+}, has the capacity to react with reactive oxygen species and generate toxic hydroxyl radicals (OH^-), consequently the Fe homeostasis systems of mammals maintain the levels of 'free' Fe in their tissues at concentrations $<10^{-23}$ M in the extracellular environment and within the range of 10^{-8}–10^{-6} M in the intracellular environment (Hider and Kong 2013) to avoid Fe toxicity. Some of the major components of these Fe homeostasis systems include the Fe transport protein transferrin (Tf) which binds Fe^{3+} with high affinity in the blood and serum, the exporter ferroportin (Fp) which actively transports Fe^{2+} out of cells, and the Fe storage protein ferritin (Ft), which converts excess intracellular soluble and reactive Fe^{2+} into insoluble and inert (e.g. less toxic) Fe^{3+} complexes (Fig. 2.1). As will be discussed later in this chapter, *Brucella* strains also have the capacity to utilize heme (Paulley et al. 2007; Ojeda 2012) as an Fe source, and thus another relevant mechanism by which mammalian hosts limit the availability of Fe to these bacteria in the extracellular environment is through the activity of the heme-binding protein hemopexin and the hemoglobin-binding protein haptoglobin, which scavenge these molecules from the serum (e.g., after leakage from damaged cells) and target them for degradation in macrophages.

The innate and acquired immune responses also have significant impacts on the availability of Fe to invading microbes, and in fact is it well documented that Fe deprivation is an important component of protective immunity in mammals (Nairz et al. 2014). During the inflammatory response, for instance, neutrophils release the Fe-binding proteins lactoferrin (Vogel 2012) and calprotectin (Nakashige et al. 2015) at the site of infection. These phagocytes also release the protein lipocalin-2,

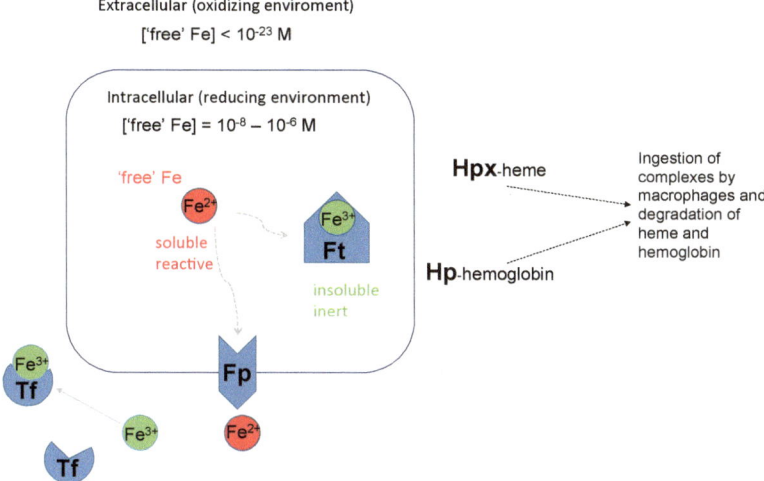

Fig. 2.1 Fe status and availability in the extracellular and intracellular environments of the mammalian host. Ft—ferritin; Fp—ferroportin; Tf—transferrin; Hpx—hemopexin; Hp—haptoglobin

also known as siderocalin, which inhibits the activity of catechol-based siderophores, small molecular weight chelators that microbes secrete into their environment to capture and transport Fe (Goetz et al. 2002; Sia et al. 2013). In addition, the liver produces the peptide hormone hepcidin (Hp), which induces the degradation of ferroportin preventing the release of Fe from the liver and spleen into circulation (Michels et al. 2015). The combined activities of lactoferrin, calprotectin and Hp thus further reduce the already low levels of free Fe available in the extracellular environment of mammals (Fig. 2.2).

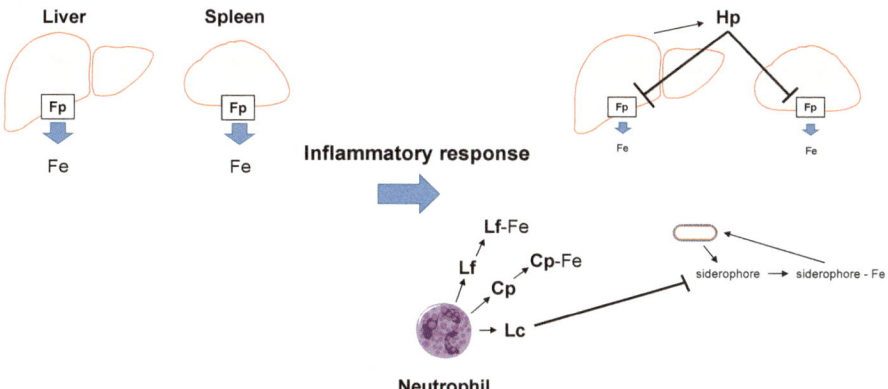

Fig. 2.2 The host inflammatory response reduces the availability of Fe in the extracellular environment. Fp–ferroportin; Hp–hepcidin; Lf–lactoferrin; Cp–calprotectin; Lc–lipocalin 2

The manner in which macrophages are 'activated' also has an impact on the availability of 'intracellular' Fe to microbes that reside within these phagocytes. This is an especially important consideration for *Brucella* strains for two reasons. First of all, the capacity of these bacteria to survive and replicate in macrophages is a critical component of their virulence (Roop et al. 2009). Secondly, it has recently been demonstrated that the activation state of host macrophages influences how well these phagocytes support the intracellular persistence of the brucellae (Xavier et al. 2013). Macrophages activated by the Th_1 cytokine interferon-γ (IFN-γ), e.g. classically-activated macrophages (CAMs), for instance, have reduced levels of Tf receptors on their surface (Byrd and Horwitz 1989), which reduces their capacity to import Fe (Fig. 2.3). The divalent cation transporter Nramp1 also pumps Fe^{2+} out of phagosomes in CAMs (Cellier et al. 2007). In addition, although hepcidin blocks the capacity of macrophages in the spleen and liver to release Fe through inhibition of Fp activity as shown in Fig. 2.2, experimental evidence indicates that CAMs have the capacity to overproduce Fp to bypass this effect and thus utilize Fp-mediated Fe release as a further mechanism for depriving intracellular pathogens of this micronutrient (Nairz et al. 2013) (Fig. 2.3). This combination of so-called 'Fe-withdrawal defenses' has been shown to be a critical component of the capacity of CAMs to control the intracellular replication of several intracellular pathogens including *Salmonella typhimurium* (Nairz et al. 2008) and *Legionella pneumophila* (Byrd and Horwitz 1989). It is well known that IFN-γ plays a crucial role in protective immunity against *Brucella* infections (Zhan and Cheers 1993; Baldwin and Goenka 2006) and that CAMs kill intracellular brucellae much better than resting macrophages (Xavier et al. 2013). But the degree to which the Fe-withdrawal defenses contribute to the brucellacidal activity of CAMs is currently unknown. Macrophages activated by the Th_2 cytokines interleukin-1 and interleukin 13 (IL-1 and IL-13) (alternatively-activated macrophages [AAMs]) are much less restrictive for the intracellular growth of *Brucella* strains than CAMs, but more importantly, recent studies have shown that AAMs provide a favorable host

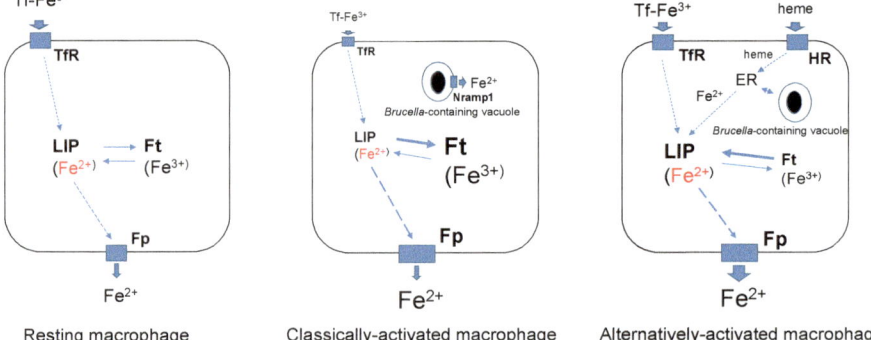

Fig. 2.3 Fe trafficking in macrophages and how macrophage activation state potentially influences the availability of this metal to the intracellular brucellae. TfR–Tf receptor; HR–heme receptor; LIP–labile iron pool; Ft–ferritin; Fp–ferroportin

cell for maintaining chronic infection in experimentally infected mice (Xavier et al. 2013). Because of their role in scavenging damaged cells and tissue components, AAMs take up a lot of Fe-and heme-containing proteins (Cairo et al. 2011). Despite the resulting increased Fe flux through AAMs, however, less of this Fe is stored in ferritin in AAMs than in CAMs, and AAMs have a larger intracellular labile Fe pools than CAMs, although increased Fp activity in AAMs prevents intracellular Fe levels from reaching toxic levels (Fig. 2.3). Whether or not the increased labile Fe pool in AAMs contributes to these phagocytes being a preferential niche for *Brucella* strains during chronic infections remains to be determined.

Three other points that warrant consideration in terms of potential Fe sources that are available to *Brucella* strains in their mammalian hosts are-(a) the recycling of Fe from erythrocytes in mammals; (b) the enhanced uptake of heme-containing proteins by AAMs; and (c) the intracellular trafficking of *Brucella* strains in host cells. The vast majority of the Fe in mammals is actively recycled by the ingestion and degradation of aged erythrocytes by macrophages in the spleen and liver (Korolnek and Hamza 2014), which are preferred organs for colonization by the brucellae in non-pregnant hosts. AAMs also ingest a considerable amount of heme-containing proteins during their role in tissue repair and remodeling (Cairo et al. 2011). The heme that is released by the degradation of hemoglobin in lysosomes during both of these processes is transported into the cytoplasm and ultimately to the endoplasmic reticulum for degradation by heme oxygenase. The Fe liberated in this fashion is incorporated into the cellular labile iron pool, stored in Ft or released from these cells by Fp for transport and redistribution to other cells (Fig. 2.3). When *Brucella* strains are ingested by host cells, effector proteins secreted by the Type IV secretion system interact with host cell proteins, which results in the *Brucella*-containing vacuoles maintaining interactions with components of the host cell endoplasmic reticulum (Celli 2015) (Fig. 2.3). As was alluded to before and will be discussed in more detail later, heme can be readily used as an Fe source by *Brucella* strains (Paulley et al. 2007; Ojeda 2012). Thus, it is intriguing that these bacteria proactively direct their intracellular trafficking in host cells toward a compartment where there is a continuous flux of a relevant Fe source.

2.3   Fe Acquisition by *Brucella*

2.3.1   Siderophore Production

Siderophores are low molecular weight chelators that bacteria and other microbes secrete into their environments under Fe-deprived conditions to capture Fe^{3+} and import it into the microbial cell using energy-dependent transport systems (Raymond and Dertz 2004). Fe^{3+} is very insoluble in aqueous environments under aerobic conditions at neutral pH due its propensity to form $Fe(OH)_3$, which has a solubility of 10^{-18} M. Thus, Fe^{3+} is only readily available under these conditions when it is bound to ligands that prevent its hydrolysis (Kosman 2013). Consequently, in order for siderophores to be effective at scavenging Fe^{3+}, these chelators must have affinities for this cation that are higher than the ligands to which it is bound in the surrounding environment. A variety of different chemical classes of siderophores have been described in the literature based on the structure of their metal binding components. Siderophores containing catechol, hydroxamate, and/or hydroxycarboxylate Fe-binding domains are found in many bacteria, but siderophores with other Fe-binding domains have also been described (Raymond and Dertz 2004). The most highly efficient siderophores contain multiple metal binding domains linked together by amino acids, polyamines or other small molecules into flexible 'scaffolds' that allow them to serve as effective Fe^{3+} chelators (Walsh and Marshall 2004).

The first siderophore described in *Brucella* was the monocatechol 2,3-dihydroxybenzoic acid (2,3-DHBA) (Fig. 2.4). López-Goñi et al. (1992) employed Fe-affinity chromatography to isolate this catechol from the supernatants of Fe-deprived cultures of *B. abortus* 2308. They also showed that 2,3-DHBA could facilitate the import of $^{55}Fe^{3+}$ into this bacterium by an energy-dependent process, and rescue this strain from Fe deprivation in bioassays. No other siderophores were identified in the supernatants from Fe-deprived cultures of 2308 or other *B. abortus* strains using this classical approach, and no known siderophores other than 2,3-DBHA or the simple monocatechols 2,3-DHBA-serine or 2,3-DHBA-glycine were found to be able to support Fe acquisition by *B. abortus* 2308 in in vitro assays. These experimental findings suggested that 2,3-DHBA might be the only siderophore produced and used by *Brucella* strains, and subsequent work by this research group provided further evidence to support this proposition (López-Goñi and Moriyón 1995).

Despite these compelling findings, and the fact that others have reported that 2,3-DHBA can serve as a siderophore in other bacteria (Hancock et al. 1977; Smith et al. 1990; Persmark et al. 1992), the ability of this simple monocatechol to serve as an 'effective' siderophore when produced endogenously by a bacterium has been called into question from a theoretical perspective (Chipperfield and Ratledge 2000) based on its 'low' Fe^{3+} binding capacity. The stability of Fe^{3+}-siderophore complexes serves as a measure of a siderophore's affinity for Fe, and is often expressed as its pFe value (Crumbliss and Harrington 2009). This value represents the negative log of the concentration of uncomplexed Fe^{3+} remaining after an equilibrium reaction between 10 μM siderophore and 1 μM Fe^{3+} at pH 7.4. The larger the pFe value, the more stable the complex formed between Fe and the siderophore under standard laboratory conditions. The reported pFe value for 2,3-DHBA is 15 (Ollinger et al. 2006), but most microbial siderophores have pFe values ranging from 18.2 to 34.3 (Crumbliss and Harrington 2009), which means they have affinities for Fe^{3+} that are orders of magnitude greater than that of 2,3-DHBA.

A possible answer to this question was provided by two subsequent studies aimed at identifying the genes involved in siderophore biosynthesis in *B. abortus* 2308. In one of these studies, a transposon mutant derived from *B. abortus* 2308 was identified that displayed increased sensitivity to the Fe-'specific' chelator

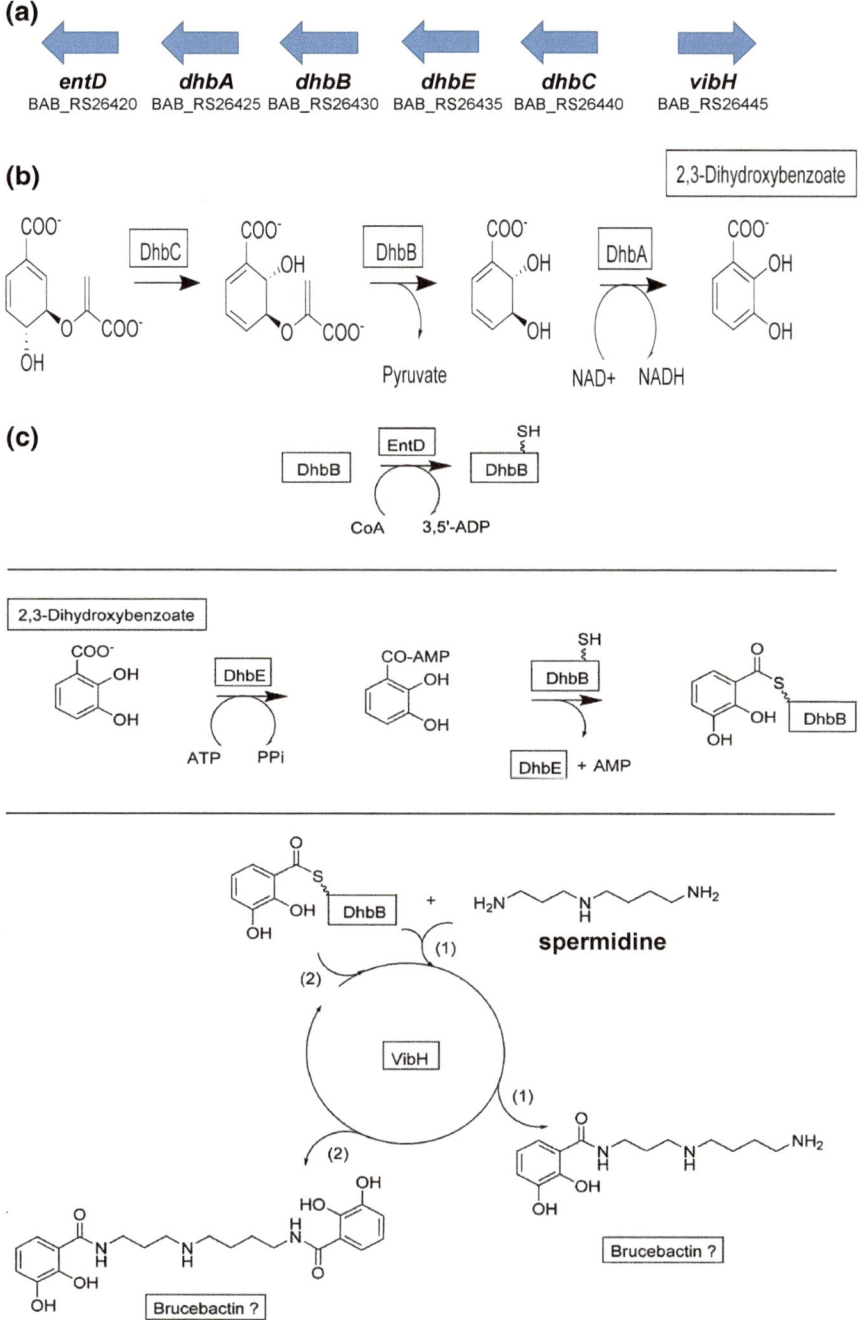

Fig. 2.4 **a** Genetic organization of the *Brucella* genes involved in brucebactin biosynthesis, **b** roles of the *dhbC*, *B* and *A* gene products in the conversion of chorismate to 2,3-dihydroxybenzoic acid (2,3-DHBA), and **c** proposed roles of the *dhbB*, *dhbE*, *entD* and *vibH* gene products in the biosynthesis of brucebactin from 2,3-DHBA and spermidine. Gene designations are those used in the *B. abortus* 2308 genome sequence in GenBank

diethylenediamine di(*o*-hydroxyphenylacetic acid) (EDDHA) compared to the parental strain, and supernatants from this mutant were found to lack a second Fe-binding catechol in addition to 2,3-DHBA that was present in 2308 (González-Carreró et al. 2003). It was also noted that the Fe acquisition defect of the *B. abortus* Tn mutant could be rescued by supplementation of cultures with the second Fe-finding catechol, but not by supplementation with 2,3-DHBA. The mTn5-disrupted gene in this mutant resides next to the genes required for 2,3-DHBA synthesis (Bellaire et al. 1999) (Fig. 2.4) and is predicted to encode a homolog of VibH, the amidase that links 2,3-DHBA to the polyamine norspermidine during vibriobactin synthesis in *V. cholerae* (Keating et al. 2002). These findings, coupled with the observation that the *B. abortus* Tn mutant produced more 2,3-DHBA than the parental strain when faced with Fe deprivation, led the authors to propose that the *Brucella* VibH homolog is involved in the conversion of 2,3-DHBA into a more complex catechol-based siderophore that has a higher affinity for Fe^{3+} than 2,3-DHBA. They proposed the name 'brucebactin' for this siderophore, but were unable to evaluate its chemical or structural features due to its reported instability. When the genes responsible for 2,3-DHBA biosynthesis in *B. abortus* 2308 were cloned and characterized by another group, it was likewise noted that these genes reside in a locus with genes homologous to those involved in the condensation of 2,3-DHBA with norspermidine during vibriobactin production (Fig. 2.4), but that the *Brucella* 2,3-DHBA/brucebactin locus notably lacks a critical gene required for the assembly of 2,3-DHBA-norspermidine complexes into vibriobactin (Bellaire et al. 2003a). This led the latter group to postulate that brucebactin is likely to be a monocatechol siderophore consisting of 2,3-DHBA linked to a small molecule such as a polyamine or an amino acid.

Based on the experimental evidence provided by the studies described in the previous paragraph, and what is known about the biosynthesis pathways typically used to assemble catechol-based siderophores in other bacteria (Walsh and Marshall 2004), it is possible to propose a biosynthetic pathway for the production of brucebactin (Fig. 2.4). The products of the *dhbC*, *B* and *A* genes convert chorismate into 2,3-DHBA in a pathway used by many bacteria to produce 2,3-DHBA as a precursor for the assembly of more complex catechol based siderophores such as the tricatechols enterobactin (Young et al. 1971; Luke and Gibson 1971) and bacillibactin (May et al. 2001). The products of the *dhbB*, *dhbE*, *entD* and *vibH* genes are then predicted to link one or two 2,3-DHBA molecules to the polyamine spermidine to produce brucebactin.

Two important theoretical considerations were taken into account in formulating the brucebactin biosynthetic pathway. The first is that *Brucella* strains do not have a gene encoding an 'EntF-type' modular non-ribosomal peptide synthetase (NRPS). This large enzyme contains adenylation, peptidyl carrier, condensation and thioesterase domains, and is required for both the activation of an amino acid and its condensation with 2,3-DHBA, and the assembly of the multiple 2,3-DHBA-small molecule complexes into complex tricatechol siderophores such as enterobactin (Walsh and Marshall 2004). Although a *Brucella* '*entF*' gene has been described previously in the literature (González-Carreró et al. 2002; Jain et al. 2011), the gene being referred to in those citations is in fact the *Brucella vibH* homolog. The second consideration has to do with the specificity of VibH itself. In *Vibrio*, this enzyme links 2,3-DHBA to norspermidine (Keating et al. 2002), but this particular polyamine is only found in a few bacteria outside of this genus. Thus, it is likely that the *Brucella* VibH homolog links 2,3-DHBA to another structurally similar polyamine such as spermidine (L. Quadri, personal communication).

A recent chemical and structural characterization of brucebactin performed in the laboratory of one of the authors of this chapter (E.S. Anderson) has provided evidence that this siderophore is indeed a dicatechol comprised of two 2,3-DHBA molecules linked by a spermidine as shown in Fig. 2.4 (Atkinson 2015). Further characterization of brucebactin will be required to determine its Fe^{3+}-binding properties, but one would predict that they are comparable to the value of 23.3 reported for the dicatechol azotochelin, which is comprised of two 2,3-DHBA molecules linked by a lysine (Cornish and Page 2000). If this is correct, it would give brucebactin an affinity for Fe^{3+} that is 8 logs higher than that of 2,3-DHBA. Perhaps more importantly, it would give the *Brucella* siderophore a pFe value in the range (e.g., >20) that has been proposed to be required for bacterial siderophores to be able to effectively compete with transferrin for Fe^{3+} in mammalian tissues (Evans et al. 2012).

2.3.2   *Fe^{3+}-Siderophore Transport Systems in Brucella*

Due to their size, Fe^{3+}-siderophore complexes require energy for their transport into the cytoplasm of bacteria. In Gram-negative bacteria, transport across the outer membrane is typically facilitated by 'gated' porins which obtain the energy they need to drive this transport from the ExbBD-TonB system (Noinaj et al. 2010). These Fe^{3+}-siderophore complexes are then bound by specific periplasmic binding proteins which direct them to cytoplasmic ABC-type permeases that mediate their passage across the cytoplasmic membrane. Once in the cytoplasm, Fe^{3+} is released from the siderophore by its reduction to $Fe^{2+}$and/or degradation of the siderophore (Cooper et al. 1978; Brickman and McIntosh 1992; Harrington and Crumbliss 2009).

Two genetic loci involved in Fe^{3+}-siderophore transport have been identified in *Brucella*—*fatBDCE* and *exbBD-tonB* (Fig. 2.5). The former is predicted to encode a transporter homologous to the periplasmic-binding-protein dependent ABC transporter that imports Fe^{3+}—anguibactin complexes across the cytoplasmic membrane in *Vibrio anguillarum* (Köster et al. 1991). As described in the previous paragraph, the latter encodes the energy transduction system required for Fe^{3+}-siderophore transport across the outer membrane (Postle and Kadner 2003). Published studies have shown that *B. abortus fatB* and *B. melitensis fatC* and *exbB* mutants cannot use brucebactin and 2,3-DHBA, respectively, as Fe sources in vitro (González-Carreró et al. 2002; Danese et al. 2004). In contrast, the identity of the genes that encode the TonB-dependent OM protein that transports Fe^{3+}-brucebactin

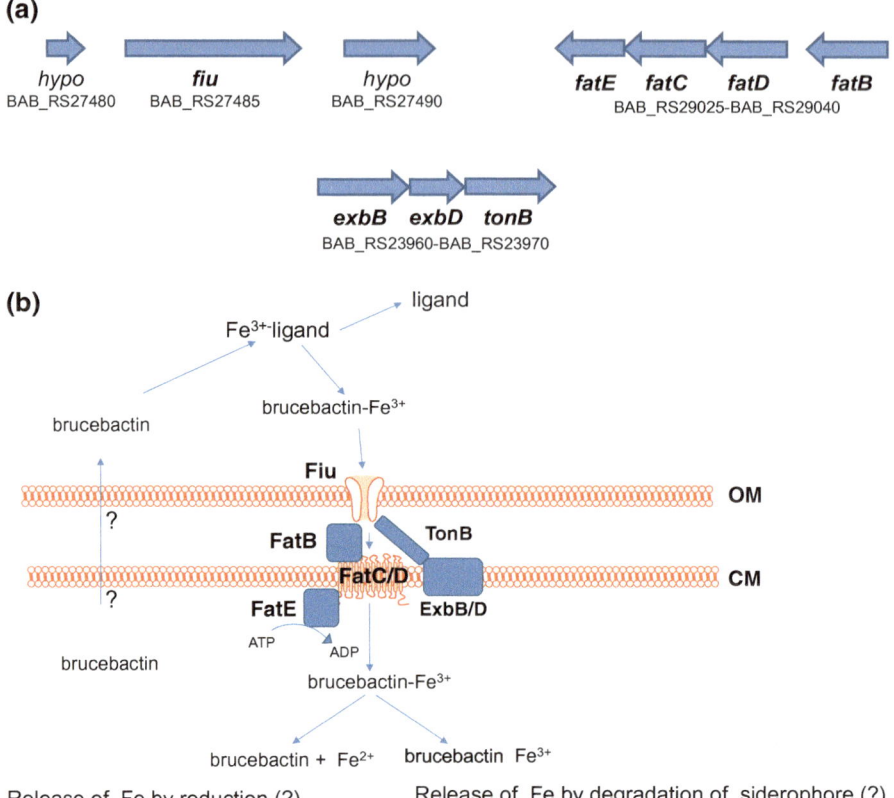

Fig. 2.5 a Genetic organization of the *Brucella fiu*, *fatBCDE* and *exbBDtonB* loci, and **b** proposed roles of the corresponding gene products in Fe^{3+}-brucebactin transport. Gene designations are those used in the *B. abortus* 2308 genome sequence in GenBank. OM–outer membrane; CM–cytoplasmic membrane

across the outer membrane is presently unclear. The two best candidates for performing this function are the *Brucella fiu* (Fig. 2.5) and *cir* (Fig. 2.6) homologs (Roop et al. 2012). Fiu and Cir are TonB-dependent outer membrane proteins that transport Fe^{3+}-2,3-DHBA and Fe^{3+}-2,3-DHBA-serine complexes into *E. coli* (Hantke 1990; Nikaido and Rosenberg 1990). However, the *Brucella cir* homolog is annotated in most in most *Brucella* genome sequences as being part of an operon encoding a BtuBCDE-type TonB-dependent cobalamin (vitamin B_{12}) transporter (Kadner 1990), and bioinformatics studies suggest that these genes are regulated by a cobalamin-dependent riboswitch (Rodionov et al. 2003). A more detailed experimental examination will be required to determine the contributions that Fiu and Cir make to Fe^{3+}-brucebactin transport in *Brucella*.

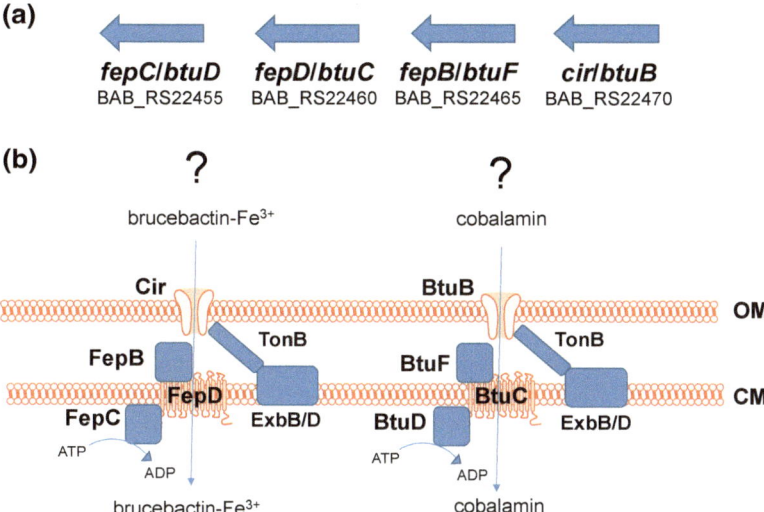

Fig. 2.6 **a** Genetic organization of the genes designated BAB_RS22455–22470 in the *Brucella abortus* 2308 genome sequence, and **b** the proposed roles of the corresponding gene products in either Fe^{3+}-brucebactin or cobalamin transport. OM–outer membrane; CM–cytoplasmic membrane

2.3.3   Siderophore Production and Virulence of Brucella abortus in Pregnant Ruminants

A *B. abortus dhbC* mutant, which cannot produce 2,3-DHBA or brucebactin, is extremely attenuated in pregnant goats (Bellaire et al. 2000) and cattle (Bellaire et al. 2003a). This mutant does not, however, display significant attenuation in BALB/c (Bellaire et al. 1999) or C57BL/6 mice (Parent et al. 2002). One explanation for these findings is the different experimental readouts used to measure virulence in these animal models. Abortion and fetal pathology, for instance, are the parameters used to measure the virulence of *Brucella* strains during experimental infections in pregnant ruminants (Bellaire et al. 2003a). Rapid and extensive proliferation of the brucellae in the placental trophoblasts surrounding the developing fetus leads to these clinical manifestations (Enright 1990), and this represents the so-called 'acute' stage of brucellosis in natural ruminant hosts (Roop et al. 2004). In contrast, the mouse model assesses the capacity of *Brucella* strains to establish and maintain 'chronic' spleen infection for weeks or months after experimental challenge (Baldwin and Winter 1994; Grilló et al. 2012), and the key host cell in these infections is the macrophage. Thus, it is tempting to speculate that siderophore production plays a more important role in Fe acquisition when *Brucella* strains are replicating in placental trophoblasts than it does when these bacteria are replicating in macrophages. The fact that *B. abortus dhbC* and *vibH* mutants exhibit little or no attenuation in cultured murine macrophages (Bellaire et al. 1999; Parent et al. 2002;

González-Carreró et al. 2002; Jain et al. 2011), and a *dhbC* mutant is not attenuated in the human monocytic cell line THP-1 (B. Bellaire, personal communication), would also seem to support this proposition. But a preliminary study has also shown that a *B. abortus dhbC* mutant is more susceptible to killing by cultured bovine macrophages than the parental 2308 strain (Bellaire 2001), suggesting that ruminant macrophages may represent a more Fe-restrictive host cell than their murine or human counterparts. Still another possibility is that siderophore production is important for Fe acquisition by the brucellae in the extracellular, rather than the intracellular, environment of the ruminant reproductive tract during pregnancy.

The documented link between siderophore production and erythritol catabolism (Bellaire et al. 2003b; Jain et al. 2011) may also explain the differential virulence properties of the *B. abortus dhbC* mutant in mice and ruminants. Ruminant placental trophoblasts produce large amounts of this 4 carbon sugar alcohol (Smith et al. 1962), while mice do not have substantial levels of this compound in their tissues. Erythritol is the preferred carbon source for *B. abortus* strains in vitro (Anderson and Smith 1965), and it has been proposed that the capacity of these bacteria to catabolize erythritol is an important virulence determinant in pregnant ruminants (Smith et al. 1962). Studies employing both *B. abortus dhbC* and *vibH* mutants have shown that *B. abortus* 2308 has a greater physiologic need for Fe when growing in the presence of erythritol than it does when growing in the presence of other carbon and energy sources, and that siderophore biosynthesis is required to meet this increased need for Fe (Bellaire et al. 2003b; Jain et al. 2011). Thus, it is conceivable that brucebactin serves as a virulence determinant for *B. abortus* strains in pregnant ruminants because its production allows the bacteria to acquire the levels of Fe they need to efficiently catabolize erythritol and support their extensive replication in the gravid ruminant reproductive tract. Clearly, there are many questions that need to be answered before we will have an accurate determination of the role that brucebactin plays in the virulence of *Brucella* strains in their natural hosts.

2.3.4 Heme as an Fe Source for Brucella Strains

By showing that exogenous heme could rescue the heme auxotrophy of a *B. abortus hemH* mutant, Almirón et al. (2001) also demonstrated that the parental 2308 strain has the capacity to transport the intact heme molecule. This was an important finding because as noted previously, the intracellular trafficking pattern of the brucellae in host macrophages places these bacteria in an environment where heme is conceivably a relevant Fe source. Subsequent studies showed that heme transport in *B. abortus* 2308 is mediated by the TonB-dependent outer membrane transporter BhuA (Paulley et al. 2007) and the periplasmic binding protein-dependent ABC transporter BhuTUV (Ojeda 2012) (Fig. 2.7). These latter studies also confirmed that *Brucella* strains can use heme as an Fe source. Moreover, the significant

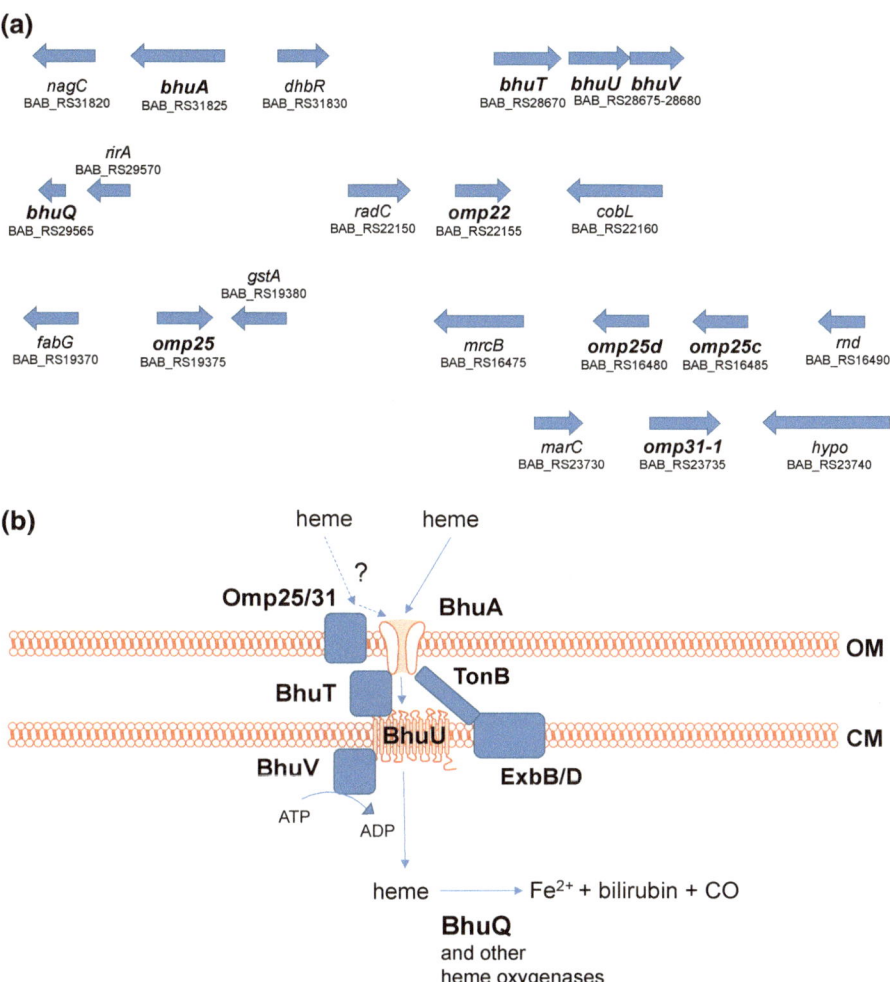

Fig. 2.7 **a** Genetic organization of the *Brucella bhuA, bhuTUV, bhuQ, omp22, omp25, omp25c omp25d,* and *omp31-1* loci, and **b** proposed roles of the corresponding gene products in heme transport and degradation. Gene designations are those used in the *B. abortus* 2308 genome sequence in GenBank. OM–outer membrane; CM–cytoplasmic membrane

attenuation displayed by a *B. abortus bhuA* mutant in cultured murine macrophages and experimentally infected mice supports the proposition that heme is indeed an important Fe source for the brucellae during intracellular replication in their mammalian hosts (Paulley et al. 2007).

The genetic organization of the *Brucella bhuA, T, U* and *V* genes is interesting. Bacterial heme transport genes typically reside in operons, which allows their coordinated expression in response to Fe and heme availability. But in *Brucella* strains, the gene encoding the outer membrane transporter (*bhuA*) resides >660 kb distant from the genes encoding the rest of the transport system (*bhuT*, *U* and *V*) (Fig. 2.7). The *bhuA* gene in *B. abortus* 2308 is also much more responsive to Fe-deprivation at the transcriptional level than the *bhuTUV* locus (Ojeda 2012). These data suggest that the *bhuA* and *bhuTV* may respond differently to relevant environmental stimuli, e.g., cellular Fe levels and the availability of heme in the external environment, but these relationships require further study. It is also not clear what biological benefit differential regulation of the *bhuA* and *bhuTUV* would provide to the brucellae.

Heme oxygenases degrade the protoporphyrin backbone of heme and release Fe^{2+} (Frankenberg-Dinkel 2004). By catalyzing this reaction, these enzymes play an important role in allowing bacteria to use internalized heme as an Fe source. But excess heme in the cytoplasm can also be toxic due of its hydrophobicity and potential to participate in the production of ROS, so heme oxygenases also play an important role in preventing heme toxicity in bacteria (Skaar et al. 2006). To date, a single heme oxygenase, BhuQ, has been identified in *Brucella* (Ojeda et al. 2012). While phenotypic analysis of a *B. abortus bhuQ* mutant clearly shows that the corresponding gene product plays a role in the capacity of these bacteria to use heme as an Fe source, these studies have also shown that the brucellae contain other, as yet unidentified, heme oxygenases that also contribute to this process (Fig. 2.7).

Another set of proteins that may contribute to the ability of *Brucella* strains to capture heme from the external environment are the Omp25/Omp31 family of proteins (Vizcaíno and Cloeckaert 2012) (Fig. 2.7). These proteins are major constituents of the outer membranes of these bacteria, and although their precise functions are unknown, experimental evidence suggests that they play important roles in the interactions of the brucellae with their mammalian hosts. One of the interesting features of the Omp25/31 proteins is that they are homologs of the so-called 'heme-binding proteins' (Hbps) that have been described in *Bartonella* (Minnick et al. 2003). In fact, the *Brucella melitensis* Omp31 protein binds heme in vitro (Delpino et al. 2006), but whether or not these proteins actually play a role in heme acquisition by *Brucella* remains to be determined.

2.3.5   Ferrous Iron Transport by Brucella Strains

The brucellae are intracellular pathogens in their mammalian hosts, and their capacity to survive and replicate efficiently in macrophages and placental trophoblasts is critical for their virulence (Roop et al. 2009). During the initial stages of their intracellular life cycle in mammalian cells these bacteria reside in acidified, membrane-bound compartments known as 'endosomal *Brucella*-containing vacuoles', or eBCVs (Starr et al. 2012). The acidic pH of the eBCVs serves as a signal that stimulates expression of the genes encoding the Type IV secretion machinery (Boschiroli et al. 2002), which translocates effector molecules into the cytoplasm of the host cell and directs the intracellular trafficking of the BCVs (de Jong et al.

2008; de Barsy et al. 2011; Marchesini et al. 2011; Salcedo et al. 2013; Myeni et al. 2013). The interactions of the eBCVs with lysosomes potentially provides the brucellae with access to Fe released into the host cell by transferrin-mediated Fe import (Anderson and Vulpe 2009), and the low pH of these intracellular compartments favors the solubility of the ferrous form of Fe (Fe^{2+}) (Kosman 2013). Recent studies have also shown that the majority of the Fe present in the so-called 'labile Fe pool' of host cells is in the form of Fe^{2+} (Hider and Kong 2013). For these reasons, it has been postulated that Fe^{2+} is a relevant Fe source for the brucellae during their intracellular replication in their mammalian hosts (Roop et al. 2012).

Brucella strains produce a single high affinity Fe^{2+} transporter (Elhassanny et al. 2013), which is predicted to be structurally and functionally similar to the FtrABCD Fe^{2+} transporter described in *Bordetella* spp. (Brickman and Armstrong 2012). Both of these transporters are specific for Fe^{2+}, and a *B. abortus ftrA* mutant is highly attenuated in experimentally infected mice (Elhassanny et al. 2013), which supports the contention that Fe^{2+} is a relevant Fe source for *Brucella* strains during infection. The *Brucella* FtrABCD transporter and its *Bordetella* counterpart are proposed to transport Fe^{2+} by the mechanism shown in Fig. 2.8. This model is based on the predicted functions of their individual components and the phenotypes of defined mutants (Brickman and Armstrong 2012; Elhassanny et al. 2013). FtrA is a member of the P19 family of periplasmic Fe-binding proteins which have well-documented roles in bacterial Fe transport (Chan et al. 2010; Koch et al. 2011). The periplasmic protein FtrB is thought to have ferroxidase activity (e.g. $Fe^{2+} \rightarrow Fe^{3+}$ activity), based on the presence of putative 'cupredoxin-like' metal binding domains (Rajasekaran et al. 2010), but this function has not been experimentally validated. FtrC is a homolog of the Fe permease Ftr1p, which has been

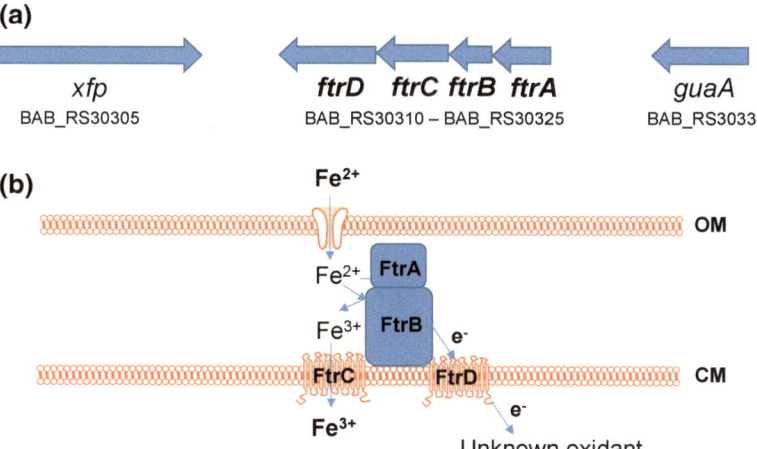

Fig. 2.8   a Genetic organization of the *Brucella ftrABCD* locus, and **b** proposed roles of the corresponding gene products in Fe^{2+} transport **b**. Gene designations are those used in the *B. abortus* 2308 genome sequence in GenBank. OM–outer membrane; CM–cytoplasmic membrane

well-studied in eukaryotic microbes (Kosman 2003). One of the distinctive features of Ftr1p is that it forms a complex with the ferroxidase Fet3p, and Ftr1p transports Fe^{2+}, but only after it has been oxidized to Fe^{3+} in a step that occurs concomitant with transport (de Silva et al. 1995; Kwok et al. 2006). FtrD shares significant amino acid homology with the ferredoxin NapH, which resides in the cytoplasmic membrane of the bacterium *Wolinella succinogenes* and is an integral component of an electron transport chain (Kern and Simon 2008). Hence, FtrD has been proposed to serve as a sink for the electrons transferred to FtrB during the oxidation of Fe^{2+}, thereby resetting the redox balance of the transporter.

It is important to note that the necessity for the coupled Fe^{2+} oxidation/permeation step depicted in Fig. 2.8 has not verified experimentally for either the *Brucella* or *Bordetella* FtrABCDs. And if it is required, it is not clear what benefit this mode of transport would provide, if any, compared to 'direct' transport of Fe^{2+}. One possibility is that converting Fe^{2+} to Fe^{3+} provides a mechanism for 'trapping' the insoluble Fe^{3+} at the site of transport. This proposition is consistent with the observation that this mechanism of Fe^{2+} transport is kinetically superior to direct Fe^{2+} transport by at least an order of magnitude in yeast (Kosman 2010). There is also evidence suggesting that this coupled oxidation/permeation step provides an indirect form of oxidative defense by converting the strong oxidant Fe^{2+} to the less reactive Fe^{3+} (Shi et al. 2003; Kosman 2010). This latter possibility is particularly intriguing when one considers that both *Brucella* and *Bordetella* rely upon a respiratory type of metabolism (Corbel and Brinley-Morgan 1984; Pittman 1984) and would be expected to be transporting Fe^{2+} in host tissues where O_2 is present.

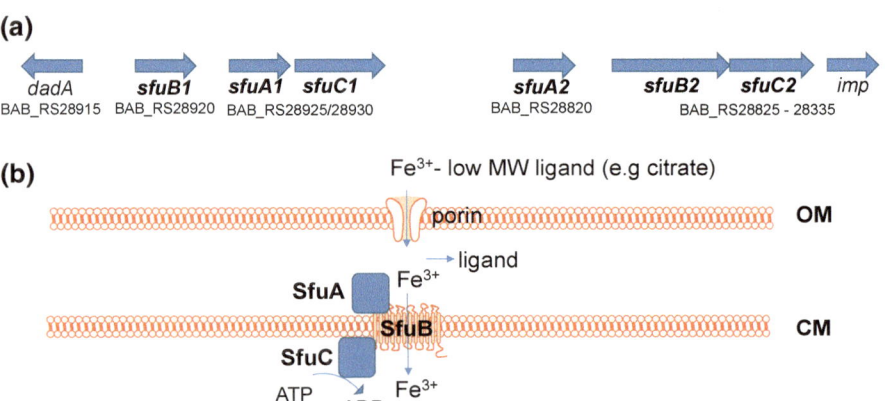

Fig. 2.9 **a** Genetic organization of the *Brucella sfu1* and *sfu2* loci, and **b** proposed roles of the corresponding gene products in Fe^{3+} transport. Gene designations are those used in the *B. abortus* 2308 genome sequence in GenBank. OM–outer membrane; CM–cytoplasmic membrane

2.3.6 Siderophore-Independent Transport of Fe^{3+} in Brucella

Fe^{3+}-siderophore complexes generally require energy for transport across the outer membrane of Gram-negative bacteria (Noinaj et al. 2010), but Fe^{3+} bound to low molecular weight chelators such as citrate can diffuse across the outer membrane via porins, where specialized Fe^{3+}-specific periplasmic protein-dependent ABC transporters can capture this Fe^{3+} and transport it across the cytoplasmic membrane. Such transporters include the Sfu (Angerer et al. 1992), Afu (Chin et al. 1996) and Yfu (Gong et al. 2001) systems that have been described in *Serratia*, *Actinobacillus* and *Yersinia*, respectively. Two sets of genes predicted to encode Sfu-type Fe^{3+} transporters have been described in *Brucella* (Jenner et al. 2009; Roop et al. 2012) (Fig. 2.9), but the roles that the corresponding gene products play in Fe transport and/or virulence are unknown.

2.4 Detoxification of Excess Intracellular Fe by *Brucella* Strains

Bacteria must not only be able to import enough Fe to meet their physiological needs, but they must also have a means of maintaining cellular levels of unincorporated Fe below 'toxic' levels. One 'indirect' mechanism that they use is to tightly regulate the genes encoding their Fe import systems so that they only import Fe when cellular levels fall below a certain threshold. How *Brucella* strains employ this strategy to prevent Fe toxicity will be discussed in detail in a subsequent section. The other two more 'direct' mechanisms that bacteria employ to prevent Fe toxicity are to (a) convert excess intracellular Fe^{2+} into a 'non-toxic' form (e.g. Fe^{3+}) for storage in proteins such as bacterioferritin and Dps; and (b) to export excess intracellular Fe^{2+} from the cell.

2.4.1 Bacterioferritin and Dps

Bacterioferritins (Bfrs) are large, 24 subunit proteins that with 12 heme groups form hollow spheres in bacterial cells (Andrews 2010). These proteins have distinctive ferroxidase centers that convert soluble Fe^{2+} to insoluble Fe^{3+} which is then stored as $2Fe(O)OH$ in the interior of these spherical proteins. Each individual Bfr can store up to 4500 atoms of Fe^{3+}, and this Fe^{3+} can be converted to Fe^{2+} and released back into the bacterial cytoplasm as needed to replenish depleted Fe levels. By converting the highly reactive Fe^{2+} to the less reactive Fe^{3+} and sequestering it away from the other components of the cytoplasm, Bfr serves not only as a depot for excess Fe but also prevents this excess Fe from reaching toxic levels (Fig. 2.10).

Fig. 2.10 **a** Genetic
organization of the *Brucella*
bfr and *dps* loci, and
b proposed roles of the
corresponding gene products
in Fe^{2+} detoxification and
storage. Gene designations
are those used in the
B. abortus 2308 genome
sequence in GenBank

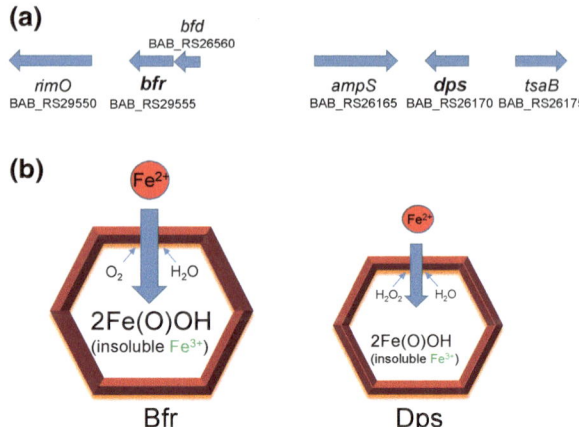

Dps is another spherical protein found in bacteria that oxidizes Fe^{2+} to Fe^{3+} and stores the insoluble Fe^{3+} in its interior (Andrews 2010) (Fig. 2.10). Although Dps, like Bfr, is considered a member of the ferritin-like superfamily of proteins, it has three important structural and functional differences when compared to Bfr. First, Dps is made up of 12 subunits, and is thus smaller than Bfr, and consequently can store less Fe (e.g., 500 atoms) than Bfr. The ferroxidase site in Dps is also different from that of Bfr, and Dps uses H_2O_2 instead of O_2 to catalyze the oxidation of Fe^{2+}. In addition, many, but not all bacterial Dps proteins, bind non-specifically to DNA. In fact, the name of this protein, Dps (DNA-binding protein from starved cells), was derived from the observation that this protein is a major nucleoid-associated protein in *E. coli* cells after they enter into stationary phase (Almirón et al. 1992). Based on these properties and the phenotypic analysis of bacterial *dps* mutants, it has been proposed that the major physiologic function of Dps is as a cellular defense against oxidative stress rather than Fe storage (Chiancone and Ceci 2010).

Brucella strains produce both Bfr (Denoel et al. 1997; Almirón and Ugalde 2010) and Dps (Kim et al. 2014) (Fig. 2.10). Phenotypic analysis of a defined mutant indicates that Bfr plays a role in Fe metabolism in *B. abortus* 2308, e.g., a *bfr* mutant is more sensitive to Fe deprivation and has reduced levels of intracellular Fe compared to the parent strain (Almirón and Ugalde 2010). The biological function of Dps, on the other hand, is presently unclear. Although the *dps* gene is strongly regulated by the general stress response sigma factor RpoE1 in *B. abortus* 2308 (Kim et al. 2014), a *B. abortus dps* mutant exhibits wild-type resistance to H_2O_2 in in vitro assays. Considering the possibility that Dps could conceivably play a compensatory role to Bfr in terms of Fe storage, it will be important to evaluate the Fe storage capabilities and virulence properties of *Brucella bfr dps* double mutants to adequately evaluate their respective biological functions.

2.4.2   *MbfA*

Another protein that has been linked to Fe detoxification in the α-proteobacteria is MbfA (Rodionov et al. 2006). This protein was initially proposed to be a membrane-bound ferritin based on the presence of conserved 'Fe-binding domains' in its N-terminus (Andrews 2010). Another distinctive feature of the MbfA proteins is that they have 'vacuolar iron transporter (VIT) domains' in their C-termini. *Agrobacterium tumefaciens* and *Bradyrhizobium japonicum mbfA* mutants are defective in Fe export and more susceptible to Fe-mediated H_2O_2 toxicity than their parental strains (Ruangkiattikul et al. 2012; Bhubhanil et al. 2014; Sankari and O'Brian 2014), which led to the proposition that MbfA serves as an Fe^{2+} exporter in these bacteria (Fig. 2.10). Consistent with their proposed role in Fe export, the *A. tumefaciens* and *B. japonicum mbfA* genes have both been shown to display increased expression in response to increasing levels of cellular Fe and decreased expression in response to Fe deprivation, and both genes are regulated by the iron response regulator Irr (Rudolph et al. 2006; Sangwan et al. 2008; Ruangkiattikul et al. 2012). Homologs of *mbfA* have also been shown to be targets of Irr repression in *Rhizobium leguminosarum* (Todd et al. 2006) and *Rhodobacter sphaeroides* (Peuser et al. 2012), but to the authors' knowledge the role of the corresponding genes products in Fe metabolism in these latter bacteria has not been examined.

The *Brucella mbfA* gene shows the same pattern of Fe-responsive expression as its *Agrobacterium* and *Bradyrhizobium* counterparts (e.g. low Fe repression, high Fe induction), and is a direct target of Irr repression in *B. abortus* 2308 (Martinson 2014). While these properties are consistent with MbfA playing a role in Fe export in *Brucella* as shown in Fig. 2.11, this function has not yet been experimentally verified. But it would certainly be important to know if Fe detoxification by Bfr and

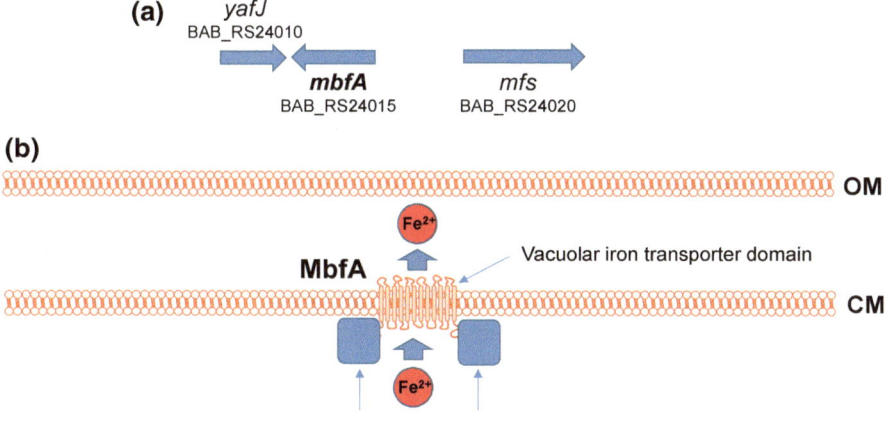

Fig. 2.11 **a** Genetic location of the *Brucella mbfA* locus, and **b** proposed role of the corresponding gene product in Fe^{2+} export **b**. Gene designations are those used in the *B. abortus* 2308 genome sequence in GenBank

MbfA, possibility in concert with Dps, plays an important role in the virulence of *Brucella* strains in their mammalian hosts. Neither *Brucella bfr* (Denoel et al. 1997; Almirón and Ugalde 2010) nor *dps* (Kim et al. 2014) mutants have been shown to be attenuated in mice or cultured mammalian cells. But it is easy to envision how the activity of MbfA could compensate for the loss of Bfr or Dps activity, if this protein is indeed an Fe^{2+} exporter.

2.5 Regulation of Fe Homeostasis in *Brucella*

Bacteria tightly regulate the expression of genes involved in Fe import, storage and export, as well as those encoding cellular proteins that require Fe for their activity. This strategy maintains Fe homeostasis, avoids Fe toxicity and sustains cellular metabolism in the face of fluctuating levels of Fe availability in their external environments. Three genetic regulators which directly control *Brucella* Fe metabolism genes, Irr, DhbR and BsrH, and one which is likely to perform this function based on its activity in other α-proteobacteria, RirA, have been described.

2.5.1 Irr

The *Brucella* iron response regulator, Irr, was originally identified by Martínez et al. (2005). Subsequent studies by this and other groups have shown that Irr is the predominant regulator of the Fe metabolism genes in these bacteria, and suggest that this regulator functions in much the same manner as the *Bradyrhizobium japonicum* Irr (Yang et al. 2006; O'Brian 2015), which is the prototype for this family of regulators. Specifically, the *Brucella* Irr is stable under conditions of Fe limitation, but is rapidly degraded under Fe-replete conditions (Martínez et al. 2005; Anderson et al. 2011). The fact that Irr can bind heme (Martínez et al. 2005) and that the Fe-responsive degradation of this protein depends upon a functional heme biosynthetic pathway (Martinson 2014) also suggests the regulatory activity of this protein is being modulated 'indirectly' by cellular Fe levels (e.g. sensing whether or not the cell has sufficient Fe to support heme biosynthesis) rather than directly sensing Fe as is the case with other bacterial Fe-responsive regulators such as Fur (Fillat 2014). Like its *B. japonicum* counterpart, the *Brucella* Irr can serve as either a transcriptional activator (Martínez et al. 2006; Anderson et al. 2011; Elhassanny et al. 2013) or repressor (Martínez et al. 2005; Ojeda et al. 2012, Martinson 2014) depending upon where it binds in the promoter region of a gene, and it activates the expression of Fe import genes and represses genes involved in Fe export and storage (Fig. 2.12).

Irr plays a critical role in the ability of *B. abortus* 2308 to maintain chronic spleen infections in experimentally infected mice (Anderson et al. 2011). This should not be too surprising since this activator is required for optimal expression of the genes encoding all three of the Fe transport systems that have been linked to

Fig. 2.12 **a** Intracellular Fe levels and the status of heme biosynthesis control the regulatory activity of the *Brucella* iron response regulator (Irr) by affecting its stability. **b** Fe homeostasis genes regulated by Irr in *B. abortus* 2308. The *solid lines* in (**b**) indicate direct interactions between Irr and the promoters of the genes depicted, the *dashed lines* indicate that it is currently unknown whether the regulatory link between Irr and these genes is direct or indirect

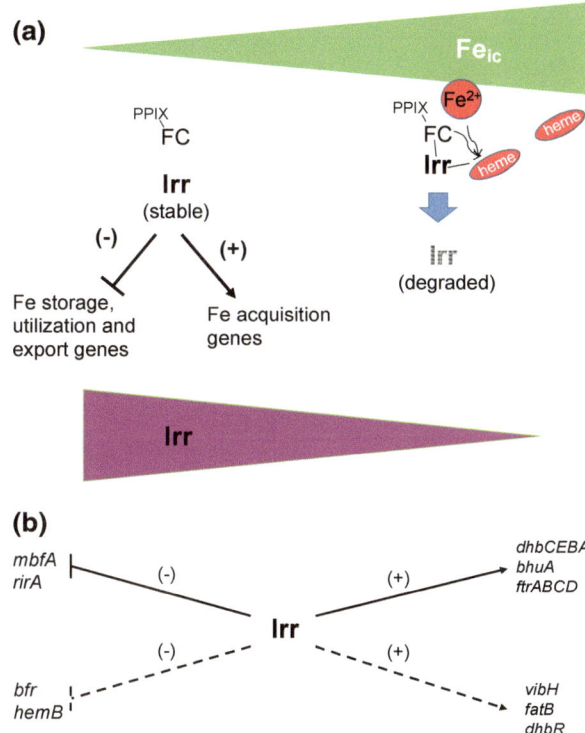

virulence in *Brucella* strains (e.g., Fe^{3+}-siderophore, heme and Fe^{2+} transport) (Martínez et al. 2006; Anderson et al. 2011; Elhassanny et al. 2013). But Irr also represses genes involved in Fe export and storage, and it would be interesting to know if down regulation of these latter genes also plays a role in this regulator's contribution to virulence.

2.5.2 *RirA*

Many of the α-proteobacteria, with the notable exception of *B. japonicum*, produce another Fe-responsive regulator known as RirA (Rodionov et al. 2006). This protein was originally identified in *Rhizobium leguminosarum* where it serves as a global repressor of Fe uptake genes during growth under Fe-replete conditions, hence its designation, rhizobial iron regulator A (Todd et al. 2002). RirA is a member of the Rrf2 family of transcriptional regulators. These regulators typically rely upon Fe-S clusters for their activity, and hence it has been proposed that like Irr, RirA does not sense Fe levels in the cell directly, but rather monitors the status of Fe-S assembly as an indirect means of sensing cellular Fe levels (Johnston et al. 2007) (Fig. 2.13).

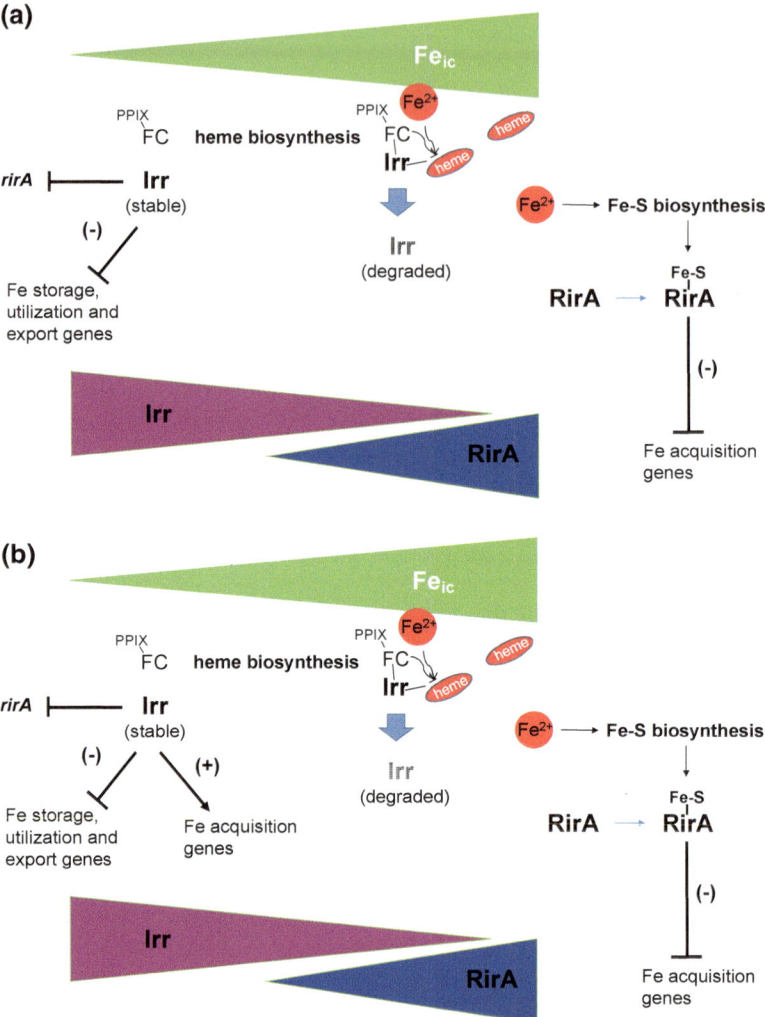

Fig. 2.13 How intracellular Fe levels and the status of heme and Fe-S center biosynthesis are proposed to balance the expression of Fe homeostasis genes in **a** *Rhizobium leguminosarum* and **b** *Agrobacterium tumefaciens* through the regulatory activities of Irr and RirA

Brucella strains produce a RirA homolog, and the corresponding gene resides in an operon with the gene encoding the heme oxygenase BhuQ (Ojeda et al. 2012). Transcription of this operon is repressed by Irr under low Fe conditions, and it lies in close proximity to the gene encoding Bfr. This genetic context and regulatory pattern is consistent with the possibility that RirA serves as an Fe-responsive regulator in *Brucella*. Microarray analysis also indicates that Fe and heme transport genes display elevated expression in a *B. abortus rirA* mutant during growth under Fe-replete conditions (Martinson 2014). Based on these observations, it is tempting to speculate that the *Brucella* Irr and RirA work together to form an integrated regulatory circuit like that proposed for the related α-proteobacterium *Agrobacterium tumefaciens* (Hibbing and Fuqua 2011), where Irr activates Fe acquisition genes and represses Fe storage, transport and utilization genes under low Fe iron conditions, and RirA represses Fe uptake genes under high Fe conditions (Fig. 2.13). But confirming this will require further experimental analysis.

2.5.3   DhbR and Other Fe Source-Specific Regulators

In addition to being up regulated in response to Fe deprivation, the genes that encode many bacterial Fe transporters are subject to a second level of regulation that ensures that they are only expressed at maximum levels if the corresponding Fe source is readily available in the external environment. For instance, the genes responsible for both biosynthesis and transport of the siderophore alcaligin in *Bordetella* are 'de-repressed' in response to Fe deprivation, but their maximum expression is dependent upon the transcriptional activator AlcR, which recognizes Fe^{3+}-alcaligin complexes imported into the cell (Brickman et al. 2001). This 'two-step' mode of regulation allows bacteria to conserve energy and prioritize the expression of their Fe acquisition genes to best fit the often changing environmental niches they inhabit (Brickman and Armstrong 2009). Similarly, the AlcR homolog DhbR is required for maximum expression of the siderophore biosynthesis genes in *B. abortus* 2308 in response to Fe deprivation (Anderson et al. 2008). Electrophoretic mobility shift assays have shown that DhbR binds directly to the *dhbC* promoter region, and genetic analysis suggests that brucebactin serves as the co-activator of DhbR (Fig. 2.14). It is presently unknown, however, if DhbR also

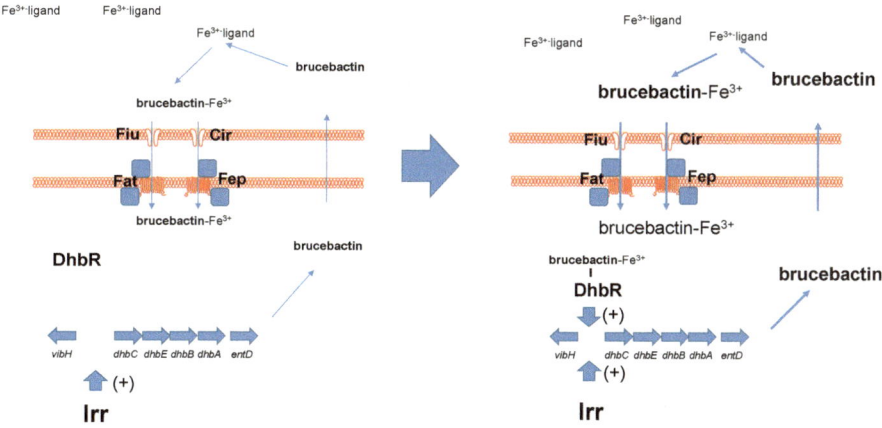

Fig. 2.14 The AraC-type transcriptional activator DhbR is required for maximum expression of the brucebactin biosynthesis genes in *B. abortus* 2308 in response to Fe deprivation. Experimental suggests that Fe^{3+}-brucebactin likely serves as a co-activator for DhbR (Anderson et al. 2008)

plays a role in the regulation of the siderophore transport genes in *Brucella*, or what role, if any, this transcriptional regulator has in the virulence of these strains in their natural ruminant hosts.

Experimental evidence likewise suggests that 'Fe source-specific' regulators play important roles in prioritizing the expression of the *Brucella* Fe^{2+} and heme transport genes. For instance, the genes encoding the Fe^{2+} transporter FtrABCD display maximum expression in *B. abortus* 2308 when this strain is subjected both Fe deprivation and exposure to acid pH (Elhassanny 2016). Irr is responsible for the low Fe-responsive activation of these genes (Elhassanny et al. 2013), but the transcriptional regulator that controls their acid-responsive induction has yet to be identified. This dual stimulus enhancement of *ftrABCD* expression may be especially beneficial to the brucellae in the endosomal BCVs in host macrophages, where they would be expected to be exposed to both Fe deprivation and acidic pH, and Fe^{2+} is likely to be a readily available Fe source (Elhassanny et al. 2013). Preliminary evidence also suggests that the genes encoding the *Brucella* heme transporters are induced in response to exposure to heme (Paulley 2007; Ojeda 2012), but the nature of this regulation has not been well-characterized.

2.5.4 BsrH

Small, untranslated regulatory RNAs (sRNAs) play critical roles in controlling gene expression in prokaryotes (Waters and Storz 2009). By facilitating or inhibiting the translation of their target mRNAs, and/or modulating the stability of these transcripts, sRNAs provide a quick mode of genetic regulation that complements and amplifies that provided by transcriptional regulators. sRNAs encoded by Fe-responsive genes are essential components of the Fe homeostasis systems of many bacteria (Massé et al. 2007). In *E. coli*, for instance, the sRNA RyhB inhibits the expression of genes encoding certain metabolic enzymes that require Fe for their activity such as the TCA cycle enzymes succinate dehydrogenase, aconitase, and fumarase A, and the Fe-cofactored superoxide dismutase SodB when the cell is growing under Fe-deprived conditions (Massé and Gottesman 2002). The *ryhB* gene is repressed by the Fe-responsive repressor Fur, and when cellular Fe levels reach a certain threshold, *ryhB* expression is repressed, and the *sdhDCAB*, *acnB*, *fumA* and *sodB* loci are efficiently expressed.

The recent discovery of the 104 nt sRNA BsrH (Peng et al. 2015) suggests that these post-transcriptional regulators may also contribute to the maintenance of Fe homeostasis in *Brucella*. The *bsrH* gene partially overlaps the gene encoding the heme biosynthesis enzyme ferrochelatase (*hemH*), but *bsrH* and *hemH* are transcribed in opposite directions. This genetic organization results in a 46 nt region of complementarity between BsrH and the *hemH* transcript, which led to the proposition that the latter gene is a target for BsrH repression. Overexpression of *bsrH* leads to reduced *hemH* expression in an *E. coli* surrogate, and produces a growth defect when *B. abortus* 2308 is grown under Fe-deprived conditions. While these findings are certainly consistent with the authors' proposition that BsrH plays a role in Fe metabolism in *Brucella* strains (Peng et al. 2015), there are several important questions that need to be addressed to determine its exact function. For instance, *bsrH* expression is reduced in response to Fe limitation in *B. abortus* 2308, and it would seem counterintuitive to relieve BsrH repression of *hemH* expression in the face of reduced cellular levels of Fe. The mechanism by which BsrH is proposed to repress *hemH* expression (e.g., as a *cis*-acting sRNA) is also atypical in comparison with known sRNAs that regulate Fe metabolism genes in other bacteria, which are *trans*-acting sRNAs that regulate genes that lie distant from the genes that encode them. Regardless, it will be important to determine precisely how sRNAs such as BsrH contribute to the regulation of Fe metabolism in *Brucella*.

2.6   Conclusions

Hopefully, this chapter will provide readers with an appreciation of how well-adapted *Brucella* strains are to defend themselves against the Fe-withdrawal defenses of their mammalian hosts. But the authors also hope that they will recognize that there is a lot more to be learned about Fe metabolism in *Brucella*. One particularly important area that needs to be addressed is how changes in Fe trafficking in the host caused by immune and other physiologic responses influences the availability of this micronutrient to *Brucella* strains in specific host cells and tissues.

Acknowledgements   The authors thank present and former members of the Roop and Almirón laboratories for their contributions to the body of work that provided the basis for preparation of this chapter. We also thank Claire Parker Siburt for preparing the siderophore biosynthesis pathways shown in Fig. 2.4. Work on *Brucella* Fe metabolism in the Roop lab has been supported by grants from the National Institute of Allergy and Infectious Disease (AI-63516) and the United States Department of Agriculture Competitive Research Grants Program (95-01995; 98-02620 and 35204-12218). Work in the Almiron lab has been supported by the Agencia Nacional de Promoción Científica y Tecnológica de la República Argentina (01-6580 and 06-00651) and the Consejo Nacional de Investigaciones Científicas y Tecnológicas de la Argentina-CONICET (PIP 5463).

References

Almirón M, Link AJ, Furlong D, Kolter R (1992) A novel DNA-binding protein with regulatory and protective roles in starved *Escherichia coli*. Gens Dev 6:2646–2654
Almirón M, Martínez M, Sanjuan N, Ugalde RA (2001) Ferrochelatase is present in *Brucella abortus* and is critical for its intracellular survival and virulence. Infect Immun 69:6225–6230
Almirón MA, Ugalde RA (2010) Iron homeostasis in *Brucella abortus*: the role of bacterioferritin. J Microbiol 48:668–673
Anderson GJ, Vulpe CD (2009) Mammalian iron transport. Cell Mol Life Sci 66:3241–3261
Anderson JD, Smith H (1965) The metabolism of erythritol by *Brucella abortus*. J Gen Microbiol 38:109–124

Anderson ES, Paulley JT, Roop RM II (2008) The AraC-like transcriptional regulator DhbR is required for maximum expression of the 2,3-dihydroxybenzoic acid biosynthesis genes in *Brucella abortus* 2308 in response to iron deprivation. J Bacteriol 190:1838–1842

Anderson ES, Paulley JT, Martinson DA, Gaines JM, Steele KH, Roop RM II (2011) The iron-responsive regulator Irr is required for wild-type expression of the gene encoding the heme transporter BhuA in *Brucella abortus* 2308. J Bacteriol 193:5359–5364

Andrews SC (2010) The ferritin-like superfamily: evolution of the biological storeman from a rubrerythrin-like ancestor. Biochim Biophys Acta 1800:691–705

Angerer A, Klupp B, Braun V (1992) Iron transport systems of *Serratia marcescens*. J Bacteriol 174:1378–1387

Archibald F (1983) *Lactobacillus plantarum*, an organism not requiring iron. FEMS Microbiol Lett 19:29–32

Atkinson XJ (2015) Determining the chemical structure of brucebactin, the sole complex siderophore utilized by the pathogenic bacterium *Brucella abortus*. Masters thesis. East Carolina University

Baldwin CL, Goenka R (2006) Host immune responses to the intracellular bacterium *Brucella*: does the bacterium instruct the host to facilitate chronic infection? Crit Rev Immunol 26:407–442

Baldwin CL, Winter AJ (1994) Macrophages and *Brucella*. In: Zwilling BS, Eisenstein TK (eds) Macrophage-pathogen interactions. Marcel Dekker, New York, NY, pp 363–380

Bellaire BH, Elzer PH, Baldwin CL, Roop RM II (1999) The siderophore 2,3-dihydoxybenzoic acid is not required for virulence of *Brucella abortus* in BALB/c mice. Infect Immun 67:2615–2618

Bellaire BH (2001) Production of the siderophore 2,3-dihydroxybenzoic acid by *Brucella abortus* is regulated independent of Fur and is required for virulence in cattle. Doctoral dissertation. Louisiana State University Health Sciences Center-Shreveport

Bellaire BH, Baldwin CL, Elzer PH, Roop RM II (2000) The siderophore 2,3-dihydroxybenzoic acid contributes to the virulence of *Brucella abortus* in ruminants. Abstr 100th Gen Meet Amer Soc Microbiol, Abstr B-17, p 44

Bellaire BH, Elzer PH, Hagius S, Walker J, Baldwin CL, Roop RM II (2003a) Genetic organization and iron-responsive regulation of the *Brucella abortus* 2,3-dihydroxybenzoic acid biosynthesis operon, a cluster of genes required for wild-type virulence in pregnant cattle. Infect Immun 71:1794–1803

Bellaire BH, Elzer PH, Baldwin CL, Roop RM II (2003b) Production of the siderophore 2,3-dihydroxybenzoic acid is required for wild-type growth of *Brucella abortus* in the presence of erythritol under low-iron conditions in vitro. Infect Immun 71:2927–2932

Bhubhanil S, Chamsing J, Sittipo P, Chaoprasid P, Sukchawalit R, Mongkolsuk S (2014) Roles of *Agrobacterium tumefaciens* membrane-bound ferritin (MbfA) in iron transport and resistance to iron under acidic conditions. Microbiology 160:863–871

Boschiroli ML, Ouahrani-Bettache S, Foulongne V, Michaux-Charachon S, Bourg G, Allardet-Servent A, Cazevieille C, Liautard JP, Ramuz M, O'Callaghan D (2002) The *Brucella suis virB* operon is induced intracellularly in macrophages. Proc Natl Acad Sci USA 99:1544–1549

Brickman TJ, Armstrong SK (2009) Temporal signaling and differential expression of *Bordetella* iron transport systems: the role of ferrimones and positive regulators. Biometals 22:33–41

Brickman TJ, Armstrong SK (2012) Iron and pH-responsive FtrABCD ferrous iron utilization system of *Bordetella* species. Mol Microbiol 86:580–593

Brickman TJ, McIntosh MA (1992) Overexpression and purification of ferric enterobactin esterase from *Escherichia coli*. Demonstration of enzymatic hydrolysis of enterobactin and its iron complex. J Biol Chem 267:12350–12355

Brickman TJ, Kang HY, Armstrong SK (2001) Transcriptional activation of *Bordetella* alcaligin siderophore genes requires the AlcR regulator with alcaligin as inducer. J Bacteriol 183:483–489

Byrd TF, Horwitz MA (1989) Interferon gamma-activated human monocytes downregulate transferrin receptors and inhibit the intracellular multiplication of *Legionella pneumophila* by limiting the availability of iron. J Exp Med 83:1457–1465

Cairo G, Recalcati S, Mantovani A, Locati M (2011) Iron trafficking and metabolism in macrophages: contribution of the polarized phenotype. Trends Immunol 32:241–247

Celli J (2015) The changing nature of the *Brucella*-containing vacuole. Cell Microbiol 17:951–958

Cellier MF, Courville P, Campion C (2007) Nramp1 phagocyte intracellular metal withdrawal defense. Microbes Infect 9:1662–1670

Chan ACK, Doukov TI, Scofield M, Tom-Yew SAL, Ramin AB, MacKichan JK, Gaynor EC, Murphy MEP (2010) Structure and function of P19, a high-affinity iron transporter of the human pathogen *Campylobacter jejuni*. J Mol Biol 401:590–604

Chiancone E, Ceci P (2010) The multifaceted capacity of Dps proteins to combat bacterial stress conditions: detoxification of iron and hydrogen peroxide and DNA binding. Biochim Biophys Acta 1800:798–805

Chin N, Frey J, Chang CF, Chang YF (1996) Identification of a locus involved in the utilization of iron by *Actinobacillus pleuropneumoniae*. FEMS Microbiol Lett 143:1–6

Chipperfield JR, Ratledge C (2000) Salicylic acid is not a bacterial siderophore: a theoretical study. Biometals 13:165–168

Cooper SR, McArdle JV, Raymond KN (1978) Siderophore electrochemistry: relation to intracellular iron release mechanism. Proc Natl Acad Sci USA 75:3551–3554

Corbel MJ, Brinley-Morgan WJ (1984) Genus *Brucella*. In: Krieg NR, Holt JG (eds) Bergey's manual of systematic bacteriology, vol 1. Williams & Wilkins, Baltimore, MD, pp 377–388

Cornish AS, Page WJ (2000) Role of molybdate and other transition metals in the accumulation of protochelin by *Azotobacter vinelandii*. Appl Environ Microbiol 66:1580–1586

Crichton R (2009) Iron metabolism—from molecular mechanisms to clinical consequences, 3rd edn. Wiley, West Sussex, UK

Crumbliss AL, Harrington JM (2009) Iron sequestration by small molecules: thermodynamic and kinetic studies of natural siderophores and synthetic model compounds. Adv Inorg Chem 61:179–251

Danese I, Haine V, Delrue R-M, Tibor A, Lestrate P, Stevaux O, Mertens P, Paquet J Y, Godfroid J, De Bolle X, Letesson JJ (2004) The Ton system, an ABC transporter, and a universally conserved GTPase are involved in iron utilization by *Brucella melitensis* 16M. Infect Immun 72:5783–5790

de Barsy M, Jamet A, Filipon D, Nicolas C, Laloux G, Rual JF, Muller A, Twizere JC, Nkengfac B, Vandenhaute J, Hill DE, Salcedo SP, Gorvel JP, Letesson JJ, De Bolle X (2011) Identification of a *Brucella* spp. secreted effector specifically interacting with human small GTPase Rab2. Cell Microbiol 13:1044–1058

de Jong MF, Sun YH, den Hartigh AB, van Dijl JM, Tsolis RM (2008) Identification of VceA and VceC, two members of the VjbR regulon that are translocated into macrophages by the *Brucella* type IV secretion system. Mol Microbiol 70:1378–1396

de Silva DM, Askwith CC, Eide D, Kaplan J (1995) The FET3 gene product required for high affinity iron transport in yeast is a cell surface ferroxidase. J Biol Chem 270:1098–1101

Delpino MV, Cassataro J, Fossati CA, Goldbaum FA, Baldi PC (2006) *Brucella* outer membrane protein Omp31 is a haemin-binding protein. Microbes Infect 8:1203–1208

Denoel PA, Crawford RM, Zygmunt MS, Tibor A, Weynants AE, Godfroid F, Hoover DL, Letesson JJ (1997) Survival of a bacterioferritin deletion mutant of *Brucella melitensis* 16M in human monocyte-derived macrophages. Infect Immun 65:4337–4340

Elhassanny AEM (2016) Characterization of FtrABCD: a ferrous iron-specific transporter that is required for the virulence of *Brucella abortus* 2308 in mice. Doctoral dissertation. East Carolina University

Elhassanny AEM, Anderson ES, Menscher EA, Roop RM II (2013) The ferrous iron transporter FtrABCD is required for the virulence of *Brucella abortus* 2308 in mice. Mol Microbiol 88:1070–1082

Enright FM (1990) The pathogenesis and pathobiology of *Brucella* infections in domestic animals. In: Nielsen KH, Duncan JR (eds) Animal brucellosis. CRC Press, Boca Raton, FL, pp 201–320

Evans RW, Kong X, Hider RC (2012) Iron mobilization from transferrin by therapeutic iron chelating agents. Biochim Biophys Acta 1820:282–290

Fillat MF (2014) The FUR (ferric uptake regulator) superfamily: diversity and versatility of key transcriptional regulators. Arch Biochem Biophys 546:41–52

Frankenberg-Dinkel N (2004) Bacterial heme oxygenases. Antioxid Redox Signal 6:825–834

Goetz DH, Holmes MA, Borregaard N, Bluhm ME, Raymond KN, Strong RK (2002) The neutrophil lipocalin NGAL is a bacteriostatic agent that interferes with siderophore-mediated iron acquisition. Mol Cell 10:1033–1043

Gong S, Bearden SW, Geoffroy VA, Fetherston JD, Perry RD (2001) Characterization of the *Yersinia pestis* Yfu ABC inorganic iron transport system. Infect Immun 67:2829–2837

González-Carreró MI, Sangari FJ, Agüero J, García-Lobo JM (2002) *Brucella abortus* 2308 produces brucebactin, a highly efficient catecholic siderophore. Microbiology 148:353–360

Grilló MJ, Blasco JM, Gorvel JP, Moriyón I, Moreno E (2012) What have we learned from brucellosis in the mouse model? Vet Res 43:e29

Hancock REW, Hantke K, Braun V (1977) Iron transport in *Escherichia coli* K-12— 2,3-dihydroxybenzoate-promoted iron uptake. Arch Microbiol 114:231–239

Hantke K (1990) Dihydroxybenzoylserine—a siderophore for *E. coli*. FEMS Microbiol Lett 67:5–8

Harrington JM, Crumbliss AL (2009) The redox hypothesis in siderophore-mediated iron uptake. Biometals 22:679–689

Hibbing ME, Fuqua C (2011) Antiparallel and interlinked control of cellular iron levels by the Irr and RirA regulators of *Agrobacterium tumefaciens*. J Bacteriol 193:3461–3472

Hider RC, Kong X (2013) Iron speciation in the cytosol: an overview. Dalton Trans 42:3220–3229

Hood MI, Skaar EP (2012) Nutritional immunity: transition metals at the pathogen-host interface. Nature Rev Microbiol 10:525–537

Jain N, Rodriguez AC, Kimsawatde G, Seleem MN, Boyle SM, Sriranganathan N (2011) Effect of *entF* deletion on iron acquisition and erythritol metabolism by *Brucella abortus* 2308. FEMS Microbiol Lett 316:1–6

Jenner DC, Dassa E, Whatmore AM, Atkins HS (2009) ABC-binding cassette systems of *Brucella*. Comp Funct Genom 2009:e354649

Johnston AWB, Todd JD, Curson AR, Lei S, Nikolaidou-Katsaridou N, Gelfand MS, Rodionov DA (2007) Living without Fur: the sublety and complexity of iron-responsive gene regulation in the symbiotic bacterium *Rhizobium* and other α-proteobacteria. Biometals 20:501–511

Kadner RJ (1990) Vitamin B_{12} transport in *Escherichia coli*: energy coupling between membranes. Mol Microbiol 4:2027–2033

Keating TA, Marshall CG, Walsh CT, Keating AE (2002) The structure of VibH represents nonribosomal peptide synthetase condensation, cyclization and epimerization domains. Nat Struct Biol 9:522–526

Kern J, Simon J (2008) Characterization of the NapGH quinol dehydrogenase complex involved in *Wolinella succinogens* nitrate respiration. Mol Microbiol 69:1137–1152

Kim H-S, Willett JW, Jain-Gupta N, Fiebig A, Crosson S (2014) The *Brucella abortus* virulence regulator, LovhK, is a sensor kinase in the general stress response signalling pathway. Mol Microbiol 94:913–925

Koch D, Chan ACK, Murphy MEP, Lilie H, Grass G, Nies DH (2011) Characterization of a dipartite iron uptake system from uropathogenic *Escherichia coli* strain F11. J Biol Chem 286:25317–25330

Korolnek T, Hamza I (2014) Like iron in the blood of the people: the requirement for heme trafficking in iron metabolism. Front Pharmacol 5:e126

Kosman DJ (2003) Molecular mechanisms of iron uptake in fungi. Mol Microbiol 47:1185–1197

Kosman DJ (2010) Redox cycling in iron uptake, efflux and trafficking. J Biol Chem 285:26729–26735

Kosman DJ (2013) Iron metabolism in aerobes: managing ferric iron hydrolysis and ferrous iron autoxidation. Coord Chem Rev 257:210–217

Köster WL, Actis LA, Waldbeser LS, Tolmasky ME, Crosa JH (1991) Molecular characterization of the iron transport system mediated by the pJM1 plasmid in *Vibrio anguillarum* 775. J Biol Chem 266:23829–23833

Kwok EY, Severance S, Kosman DJ (2006) Evidence for iron channeling in the Fet3p-Ftr1p high-affinity iron uptake complex in the yeast plasma membrane. Biochemistry 45:6317–6327

Lengeler JW, Drews G, Schlegel HG (1999) Biology of the prokaryotes. Georg Thieme Verlag, Stuttgart, Germany

López-Goñi I, Moriyón I (1995) Production of 2,3-dihydroxybenzoic acid by *Brucella* species. Curr Microbiol 31:291–293

López-Goñi I, Moriyón I, Neilands JB (1992) Identification of 2,3-dihydroxybenzoic acid as a *Brucella abortus* siderophore. Infect Immun 60:4496–4503

Luke RKJ, Gibson F (1971) Location of three genes concerned with the conversion of 2,3-dihydroxybenzoate into enterochelin in *Escherichia coli* K-12. J Bacteriol 107:557–562

Marchesini MI, Herrmann CK, Salcedo SP, Gorvel JP, Comerci DJ (2011) In search of *Brucella abortus* type IV secretion substrates: screening and identification of four proteins translocated into host cells through VirB system. Cell Microbiol 13:1261–1274

Martínez M, Ugalde RA, Almirón M (2005) Dimeric *Brucella abortus* Irr protein controls its own expression and binds haem. Microbiology 151:3427–3433

Martínez M, Ugalde RA, Almirón M (2006) Irr regulates brucebactin and 2,3-dihydroxybenzoic acid biosynthesis, and is implicated in the oxidative stress resistance and intracellular survival of *Brucella abortus*. Microbiology 152:2591–2598

Martinson DA (2014) The iron response regulator controls iron homeostasis in *Brucella*. Doctoral dissertation. East Carolina University

Massé E, Gottesman S (2002) A small RNA regulates the expression of genes involved in iron metabolism in *Escherichia coli*. Proc Natl Acad Sci USA 99:4620–4625

Massé E, Salvail H, Desnoyers G, Arguin M (2007) Small RNAs controlling iron metabolism. Curr Opin Microbiol 10:140–145

May JJ, Wendrich MT, Marahiel MA (2001) The *dhb* operon of *Bacillus subtilis* encodes the biosynthetic template for the catecholic siderophore 2,3-dihydroxybenzoate-glycine-threonine trimeric ester bacillibactin. J Biol Chem 276:7209–7217

Michels K, Nemeth E, Ganz T, Mehrad B (2015) Hepcidin and host defense against infectious diseases. PLoS Pathog 11:e1004998

Minnick MF, Sappington KN, Smitherman LS, Andersson SG, Karlberg O, Carroll JA (2003) Five-member gene family of *Bartonella quintana*. Infect Immun 71:814–821

Myeni S, Child R, Ng TW, Kupko JJ III, Wehrly TD, Porcella SF, Knodler LA, Celli J (2013) *Brucella* modulates secretory trafficking via multiple Type IV secretion effector proteins. PLoS Pathog 9:e1003556

Nairz M, Fritsche G, Brunner P, Talasz H, Hantke K, Weiss G (2008) Interferon-γ limits the availability of iron for intramacrophage *Salmonella typhimurium*. Eur J Immunol 38:1923–1936

Nairz M, Schleicher U, Schroll A, Sonnweber T, Theurl I, Ludwiczek S, Talasz H, Brandacher G, Moser PL, Muckenthaler MU, Fang FC, Bogdan C, Weiss G (2013) Nitric oxide-mediated regulation of ferroportin-1 controls macrophage iron homeostasis and immune function in *Salmonella* infection. J Exp Med 210:855–873

Nairz M, Haschka D, Demetz E, Weiss G (2014) Iron at the interface of immunity and infection. Front Pharmacol 5:e152

Nakashige TG, Zhang B, Krebs C, Nolan EM (2015) Human calprotectin is an iron-sequestering host-defense protein. Nature Chem Biol 11:765–771

Nikaido H, Rosenberg EY (1990) Cir and Fiu proteins in the outer membrane of *Escherichia coli* catalyze transport of monomeric catechols: study with & β-lactam antibiotics containing catechol and analogous groups. J Bacteriol 172:1361–1367

Noinaj N, Guillier M, Barnard TJ, Buchanan SK (2010) TonB-dependent transporters: regulation, structure and function. Ann Rev Microbiol 64:43–60

O'Brian MR (2015) Perception and homeostatic control of iron in the rhizobia and related bacteria. Annu Rev Microbiol 69:229–245

Ojeda JF (2012) The *bhuTUV* and *bhuO* genes play vital roles in the ability of *Brucella abortus* to use heme as an iron source and are regulated in an iron-responsive manner by RirA and Irr. Doctoral dissertation. East Carolina University

Ojeda JF, Martinson DA, Menscher EA, Roop RM II (2012) The *bhuQ* gene encodes a heme oxygenase that contributes to the ability of *Brucella abortus* 2308 to use heme as an iron source and is regulated by Irr. J Bacteriol 194:4052–4058

Ollinger J, Song KB, Antelmann H, Hecker M, Helmann JD (2006) Role of the Fur regulon in iron transport in *Bacillus subtilis*. J Bacteriol 188:3664–3673

Parent MA, Bellaire BH, Murphy EA, Roop RM II, Elzer PH, Baldwin CL (2002) *Brucella abortus* siderophore 2,3-dihydroxybenzoic acid (DHBA) facilitates intracellular survival of the bacteria. Microb Pathogen 32:239–248

Paulley JT (2007) Production of BhuA by *B. abortus* is required for hemin utilization and virulence and is dependent on the transcriptional regulators RirA and ChrA. Doctoral dissertation. East Carolina University

Paulley JT, Anderson ES, Roop RM II (2007) *Brucella abortus* requires the heme transporter BhuA for maintenance of chronic infection in BALB/c mice. Infect Immun 75:5248–5254

Peng X, Dong H, Wu Q (2015) A new *cis*-encoded sRNA, BsrH, regulating the expression of *hemH* gene in *Brucella abortus* 2308. FEMS Microbiol Lett 362:1–7

Persmark MD, Expert D, Neilands JB (1992) Ferric iron uptake in *Erwinia chrysanthemi* mediated by chrysobactin and related catechol-type compounds. J Bacteriol 174:4783–4789

Peuser V, Remes B, Klug G (2012) Role of the Irr protein in the regulation of iron metabolism in *Rhodobacter sphaeroides*. PLoS One 7:e42231

Pittman M (1984). Genus *Bordetella*. In: Krieg NR, Holt JG (eds) Bergey's manual of systematic bacteriology, vol 1. Williams & Wilkins, Baltimore, MD, pp 388–393

Posey JE, Gherardini FC (2000) Lack of a role for iron in the Lyme disease pathogen. Science 288:1651–1653

Postle K, Kadner RJ (2003) Touch and go: tying TonB to transport. Mol Microbiol 49:869–882

Rajasekaran MB, Nilapwar S, Andrews SC, Watson KA (2010) EfeO-cupredoxins: major new members of the cupredoxin superfamily with roles in bacterial iron transport. Biometals 23:1–17

Raymond KN, Dertz EA (2004) Biochemical and physical properties of siderophores. In: Crosa JH, Mey AR, Payne SM (eds) Iron transport in bacteria. ASM Press, Washington, DC, pp 3–17

Rodionov DA, Vitreschak AG, Mironov AA, Gelfand MS (2003) Comparative genomics of the vitamin B_{12} metabolism and regulation in prokaryotes. J Biol Chem 278:41148–41159

Rodionov DA, Gelfand MS, Todd JD, Curson ARJ, Johnston AWB (2006) Computational reconstruction of iron—and manganese-responsive transcriptional networks in α-proteobacteria. PLOS Comput Biol 2:e163

Roop RM II, Bellaire BH, Valderas MW, Cardelli JA (2004) Adaptation of the brucellae to their intracellular niche. Mol Microbiol 52:621–630

Roop RM II, Gaines JM, Anderson ES, Caswell CC, Martin DW (2009) Survival of the fittest: how *Brucella* strains adapt to their intracellular niche in the host. Med Microbiol Immunol 198:221–238

Roop RM II, Anderson E, Ojeda J, Martinson D, Menscher E, Martin DW (2012) Metal acquisition by *Brucella* strains. In: López-Goñi I, O'Callaghan D (eds) *Brucella*—molecular microbiology and genomics. Caister Academic Press, Norfolk, UK, pp 179–199

Ruangkiattikul N, Bhubhanil S, Chamsing J, Niamyim P, Sukchawalit R, Mongkolsuk S (2012) *Agrobacterium tumefaciens* membrane-bound ferritin plays a role in protection against hydrogen peroxide toxicity and is negatively regulated by the iron response regulator. FEMS Microbiol Lett 329:87–92

Rudolph G, Semini G, Hauser F, Lindemann A, Friberg M, Hennecke H, Fischer HM (2006) The iron control element acting in positive and negative control of iron-regulated *Bradyrhizobium japonicum* genes, is a target of the Irr protein. J Bacteriol 188:733–744

Salcedo SP, Marchesini MI, Degos C, Terwagne M, von Bargen K, Lepidi H, Herrmann CK, Santos Lacerda TL, Imbert PRC, Pierre P, Alexopoulo L, Letesson JJ, Comerci DJ, Gorvel JP (2013) BtpB, a novel TIR-containing effector protein with immune modulatory functions. Front Cell Infect Microbiol 3:e28

Sangwan I, Small SK, O'Brian MR (2008) The *Bradyrhizobium japonicum* Irr protein is a transcriptional repressor with high-affinity DNA-binding activity. J Bacteriol 190:5172–5177

Sankari S, O'Brian MR (2014) A bacterial iron exporter for maintenance of iron homeostasis. J Biol Chem 289:16498–16507

Shi X, Stoj C, Romeo A, Kosman DJ, Zhu Z (2003) Fre1p Cu^{2+} reduction and Fet3p Cu^{1+} oxidation modulate copper toxicity in *Saccharomyces cerevisiae*. J Biol Chem 278:50309–50315

Sia AK, Allred BE, Raymond KN (2013) Siderocalins: siderophore binding proteins evolved for primary pathogen host defense. Curr Opin Chem Biol 17:150–157

Skaar EP, Gaspar AH, Schneewind O (2006) *Bacillus anthracis* IsdG, a heme-degrading monooxygenase. J Bacteriol 188:1071–1080

Smith AW, Freeman S, Minnett WG, Lambert PA (1990) Characterisation of a siderophore from *Acinetobacter calcoaceticus*. FEMS Microbiol Lett 70:29–32

Smith H, Williams AE, Pearce JH, Keppie J, Harris-Smith PW, Fitz-George RB, Witt K (1962) Foetal erythritol: a cause of the localization of *Brucella abortus* in bovine contagious abortion. Nature 193:47–49

Starr T, Child R, Wehrly TD, Hansen B, Hwang S, López-Otin C, Virgin HW, Celli J (2012) Selective subversion of autophagy complexes facilitates completion of the *Brucella* intracellular cycle. Cell Host Microbe 11:33–45

Todd JD, Wexler M, Sawers G, Yeoman KH, Poole PS, Johnston AWB (2002) RirA, an iron-responsive regulator in the symbiotic bacterium *Rhizobium leguminosarum*. Microbiology 148:4059–4071

Todd JD, Sawers G, Rodionov DA, Johnston AWB (2006) The *Rhizobium leguminosarum* regulator IrrA affects the transcription of a wide range of genes in response to Fe availability. Mol Gen Genomics 275:564–577

Vizcaíno N, Cloeckaert A (2012) Biology and genetics of the *Brucella* outer membrane. In: López-Goñi I, O'Callaghan D (eds) *Brucella*—molecular microbiology and genomics. Caister Academic Press, Norfolk, UK, pp 133–161

Vogel HJ (2012) Lactoferrin, a bird's eye view. Biochem Cell Biol 90:233–244

Walsh CT, Marshall CG (2004) Siderophore biosynthesis in bacteria. In: Crosa JH, Mey AR, Payne SM (eds) Iron transport in bacteria. ASM Press, Washington, DC, pp 18–37

Waring WS, Elberg SS, Schneider P, Green W (1953) The role of iron in the biology of *Brucella suis* I. Growth and nutrition. J Bacteriol 66:82–91

Waters LS, Storz G (2009) Regulatory RNAs in bacteria. Cell 136:615–628

Xavier MN, Winter MG, Spees AM, den Hartigh AB, Nguyen K, Roux CM, Silva TMA, Atluri VL, Kerrinnes T, Keestra AM, Monack DM, Luciw PA, Eigenheer RA, Bäumler AJ, Santos RL, Tsolis RM (2013) PPARγ-mediated increase in glucose availability sustains chronic *Brucella abortus* infection in alternatively activated macrophages. Cell Host Microbe 14:159–170

Yang J, Sangwan I, Lindemann A, Hauser F, Hennecke H, Fischer HM, O'Brian MR (2006) *Bradyrhizobium japonicum* senses iron through the status of haem to regulate iron homeostasis and metabolism. Mol Microbiol 60:427–437

Young IG, Langman L, Luke RKJ, Gibson F (1971) Biosynthesis of the iron-transport compound enterochelin: mutants of *Escherichia coli* unable to synthesize 2,3-dihydroxybenzoate. J Bacteriol 106:51–57

Zhan Y, Cheers C (1993) Endogenous gamma interferon mediates resistance to *Brucella abortus* infection. Infect Immun 61:4899–4901

Chapter 3
Manganese

R. Martin Roop II, Joshua E. Pitzer, John E. Baumgartner
and Daniel W. Martin

Abstract *Brucella* strains produce a single high affinity Mn transporter, MntH, and expression of the corresponding gene is regulated by the Mn-responsive transcriptional regulator Mur. Phenotypic analysis of a variety of mutants has shown that Mn-dependent enzymes play a crucial role in the basic physiology and virulence of *Brucella* strains, and suggest that these bacteria may also be modulate their cellular Mn content as a mechanism to resist oxidative stress.

Keywords *Brucella* · Manganese · SodA · PykM

3.1 Manganese as a Micronutrient and Bacterial Mn Transport

Many bacterial proteins require Mn for their activity. For instance, Mn-dependent superoxide dismutases are important antioxidants in bacteria, and some isoforms of the key metabolic enzymes pyruvate kinase (Hohle and O'Brian 2012) and ribonucleotide reductase (Cotruvo and Stubbe 2010) require Mn for their activities. Mn is an integral component of photosynthetic apparatus of the cyanobacteria (Young et al. 2015), and some of the enzymatic activities that that control the intracellular levels of signaling molecules such as (p)ppGpp (De Boer et al. 1977; Heinemeyer et al. 1978; Johnson et al. 1979) and c-diGMP (Bobrov et al. 2005; Tamayo et al. 2005) are also dependent upon this metal. Some bacteria, like *Lactobacillus plantarum*, *Borrelia burgdorferi* and *Bradyrhizobium japonicum* require Mn as an essential micronutrient during all phases of growth (Archibald and Duong 1984; Ouyang et al. 2009; Hohle and O'Brian 2012). Others, like *Escherichia coli*, only appear to need Mn during periods of oxidative stress (Anjem et al. 2009), when this bacterium actively transports Mn and substitutes it for Fe in a

R.M. Roop II (✉) · J.E. Pitzer · J.E. Baumgartner · D.W. Martin
Department of Microbiology and Immunology, Brody School of Medicine,
East Carolina University, Greenville, NC 27834, USA
e-mail: roopr@ecu.edu

© Springer International Publishing AG 2017   41
R. Martin Roop II and Clayton C. Caswell (eds.), *Metals and the Biology
and Virulence of Brucella*, DOI 10.1007/978-3-319-53622-4_3

subset of its proteins to protect them from oxidative damage resulting from intra-cellular Fenton chemistry.

Two major types of transporters are responsible for high affinity Mn transport in most bacteria that have been studied (Papp-Wallace and Maguire 2006). One is the Sit-type transporters, which represent a subclass of the Mn/Fe/Zn-specific cluster A-I ABC transporters (Berntsson et al. 2010; Lisher and Giedroc 2013). These trans-porters consist of a periplasmic binding protein, a permease which transports the metal across the cytoplasmic membrane, and an ATPase which provides energy for this transport. Some of these transporters such as the SitABCD in *Sinorhizobium meliloti*, are specific for Mn (Platero et al. 2003), but others like the YfeABCD system in *Yersinia pestis*, can transport either Mn or Fe (Bearden and Perry 1999). The second type of high affinity Mn transporter that is widespread in bacteria is the proton sym-porter MntH. This protein is an ortholog of the natural resistance-associated macro-phage protein (Nramp)-type transporters that transport a variety of cations across membranes in eukaryotic cells (Vidal et al. 1993; Cellier et al. 1994; Gunshin et al. 1997). Although initial studies with the *E. coli* MntH suggested that this protein might transport Mn, Fe and other divalent cations (Makui et al. 2000), subsequent studies have shown that this and other bacterial MntH proteins are Mn-specific (Kehres et al. 2000; Papp-Wallace and Maguire 2006). Some bacteria such as *Salmonella typhi-murium* (Kehres et al. 2002a, b), *Staphylococcus aureus* (Horsburgh et al. 2002) and *Bacillus subtilis* (Que et al. 2000) possess both Sit- and MntH-type Mn transporters, and studies suggest that they preferentially utilize these transporters under different environmental conditions. For instance, optimal Mn transport mediated by MntH in *S. typhimurium* occurs at pH 6 and below, while the SitABC system in this bacterium displays its optimal activity at pH 7 and above (Kehres et al. 2002a, b).

3.2 MntH-Mediated Mn Transport Is Critical for the Basic Biology and Virulence of *Brucella* Strains

Early studies showed that Mn is important for the growth of *Brucella* strains in vitro (Evenson and Gerhardt 1955). MntH is the only high affinity Mn^{2+} transporter produced by *Brucella* strains (Anderson et al. 2009) (Fig. 3.1), and the phenotype displayed by a *B. abortus mntH* mutant indicates that Mn is in fact an essential micronutrient for these bacteria during all stages of growth. The colonies produced by the *mntH* mutant on nutritionally-replete solid media, for instance, are much smaller and take longer to develop than those produced by the parental 2308 strain (Fig. 3.2). This mutant also exhibits an extended generation time compared to 2308 during exponential growth in either a Mn-deprived medium (Fig. 3.2) or brucella broth (Anderson et al. 2009). This delayed growth can be alleviated by culturing the *mntH* mutant on solid media or in broth supplemented with increased levels of Mn (Fig. 3.2) (Anderson et al. 2009). This basal requirement for Mn appears to be similar to that described for *Bradyrhizobium japonicum*, where the Mn-dependent enzymes PykM and SodM play crucial roles in this bacterium's metabolism (Hohle

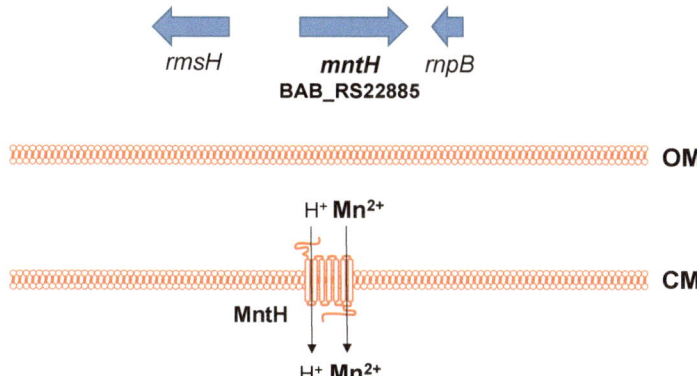

Fig. 3.1 Role of the *Brucella* MntH in Mn^{2+} transport. The corresponding gene designations are those used in the *B. abortus* 2308 genome sequence. *OM* Outer membrane; *CM* Cytoplasmic membrane

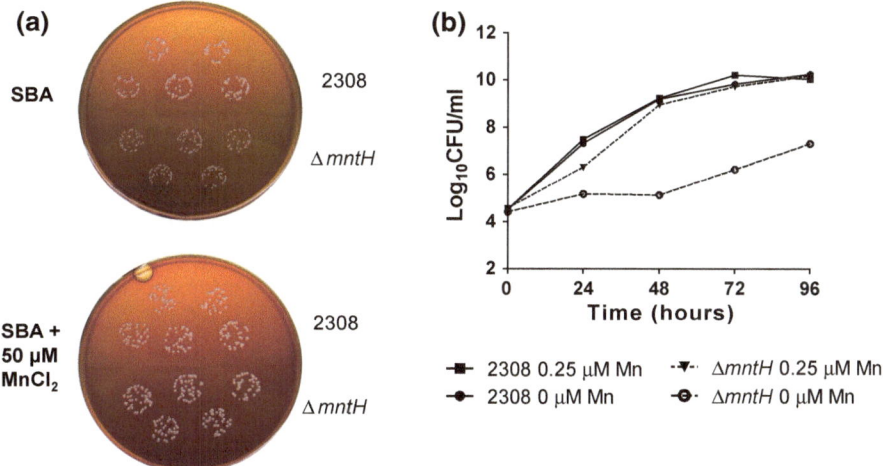

Fig. 3.2 **a** Growth of *B. abortus* 2308 and an isogenic *mntH* mutant following 48 h incubation on Schaedler agar supplemented with 5% bovine blood with or without the addition of 50 μM $MnCl_2$, and **b** growth of these strains a low Mn minimal medium with or without 0.25 μM $MgCl_2$

and O'Brian 2012), and similar relationships between Mn and wild-type growth of *Sinorhizobium meliloti* (Davies and Walker 2007) and *Agrobacterium tumefaciens* (Heindl et al. 2016) have also been reported. Thus, it is possible that reliance upon Mn as an essential micronutrient might be a common, if not universal, feature of the α-proteobacteria in general.

Limiting the availability of Mn to invading pathogens is a critical component of the mammalian host defense against microbial infections (Kehl-Fie and Skaar 2012). During the inflammatory response, for example, neutrophils release the

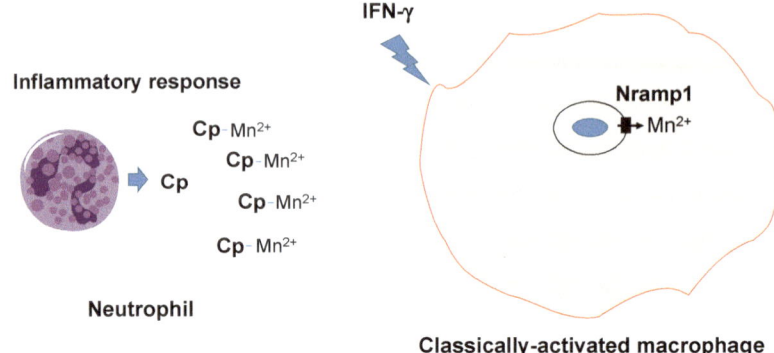

Fig. 3.3 Mn-deprivation defenses employed by neutrophils and classically-activated macrophages to limit the replication of invading pathogens. *Cp* Calprotectin

protein calprotectin (Corbin et al. 2008) which chelates Mn^{2+} in the extracellular environment (Fig. 3.3) and prevents its acquisition by bacterial pathogens such as *Staphylococcus aureus* (Kehl-Fie et al. 2013). The activation of infected macrophages by INF-γ also leads to Nramp1-mediated efflux of Mn^{2+} out of microbe-containing phagosomes (Fig. 3.3) within these phagocytes (Cellier et al. 2007), and Nramp1-mediated Mn deprivation has been shown to be an important host defense against intracellular pathogens such as *Salmonella typhimurium* (Zaharik et al. 2004). High affinity Mn transporters allow bacterial pathogens to counter these host defenses (Janulczyk et al. 2003; McAllister et al. 2004; Zaharik et al. 2004; Ouyang et al. 2009; Kehl-Fie et al. 2013; Perry et al. 2012; Crump et al. 2014), and indeed, MntH is an essential virulence determinant for *B. abortus* 2308 (Anderson et al. 2009). *Brucella* strains are predominantly intracellular pathogens and their capacity to persist in host macrophages is critical for their virulence (Roop et al. 2009). Thus, the link between MntH and virulence in *B. abortus* 2308 is not too surprising. But what is somewhat remarkable is the severity of the attenuation displayed by the *B. abortus mntH* mutant in C57BL/6 mice (Fig. 3.4)

Fig. 3.4 Spleen colonization profiles of *B. abortus* 2308 (*white bars*), an isogenic *mntH* mutant (*black bars*) and a complemented *mntH* mutant (*gray bars*) in C57BL/6 mice. Mice were infected with 5×10^4 brucellae via the intraperitoneal route

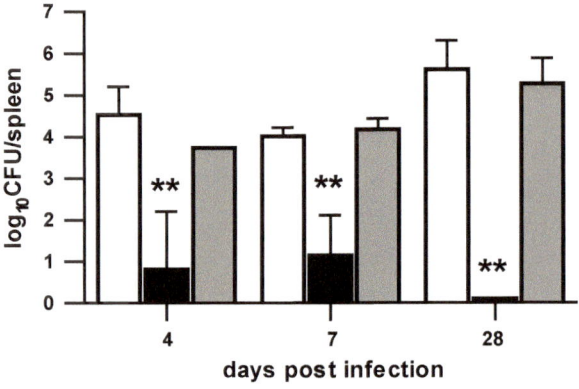

(Anderson et al. 2009). This mouse strain lacks a functional Nramp1 (Malo et al. 1994; Arpaia et al. 2011), and other bacterial Mn transport mutants often only exhibit attenuation in Nramp-positive mouse strains (Zaharik et al. 2004) and cell lines (Champion et al. 2011). Moreover, the attenuation displayed by the *B. abortus mntH* mutant is not intensified when its virulence is evaluated in a derivative of C57BL/6 mice containing a functional Nramp1 (Arpaia et al. 2011) (data not shown). These experimental findings suggest that either the basal levels of Mn present in host tissues are insufficient to sustain the persistence of *Brucella* strains in the absence of a high affinity Mn transporter, or that upon the induction of the inflammatory response, the activity of calprotectin alone is sufficient to reduce the levels of available Mn in host tissues to these levels. Thus, it will be important to determine what contribution the Mn-chelating activity of calprotectin plays in the host defense against *Brucella* infections.

3.3 Mn-Dependent Proteins that Contribute to the Virulence of *Brucella* Strains

The extreme attenuation displayed by the *mntH* mutant in mice indicates that Mn-dependent proteins likely play critical roles in the virulence of *Brucella* strains. One such protein has been identified, and several other likely candidates are described below.

3.3.1 SodA

As their name implies, superoxide dismutases (SODs) are enzymes that detoxify the superoxide anion (O_2^-), and they serve as important antioxidants for many bacteria (Imlay 2008). *Brucella* strains produce two forms of SOD, a Mn co-factored SOD, SodA, which resides in the cytoplasm (Sriranganathan et al. 1991; Stabel et al. 1994), and a Cu/Zn co-factored SOD, SodC, which resides in the periplasm (Beck et al. 1990; Bricker et al. 1990; Stabel et al. 1994). The charged nature of O_2^- prevents it from freely diffusing across the cytoplasmic membrane of bacterial cells (Hassan and Fridovich 1979), and accordingly, SodA protects the brucellae from endogenous O_2^- generated by their respiratory metabolism (Martin et al. 2012), while SodC protects these bacteria from exogenous O_2^- generated by the respiratory burst of host phagocytes (Gee et al. 2005) (Fig. 3.5a).

One important role that has been proposed for high affinity Mn transporters is to provide bacterial cells with sufficient levels of this metal to support the activity of Mn co-factored SODs (Kehl-Fie et al. 2011; Hohle and O'Brian 2012), and experimental evidence supports this proposition in *Brucella* (Martin et al. 2012). A *B. abortus mntH* mutant, for instance, produces significantly lower levels of

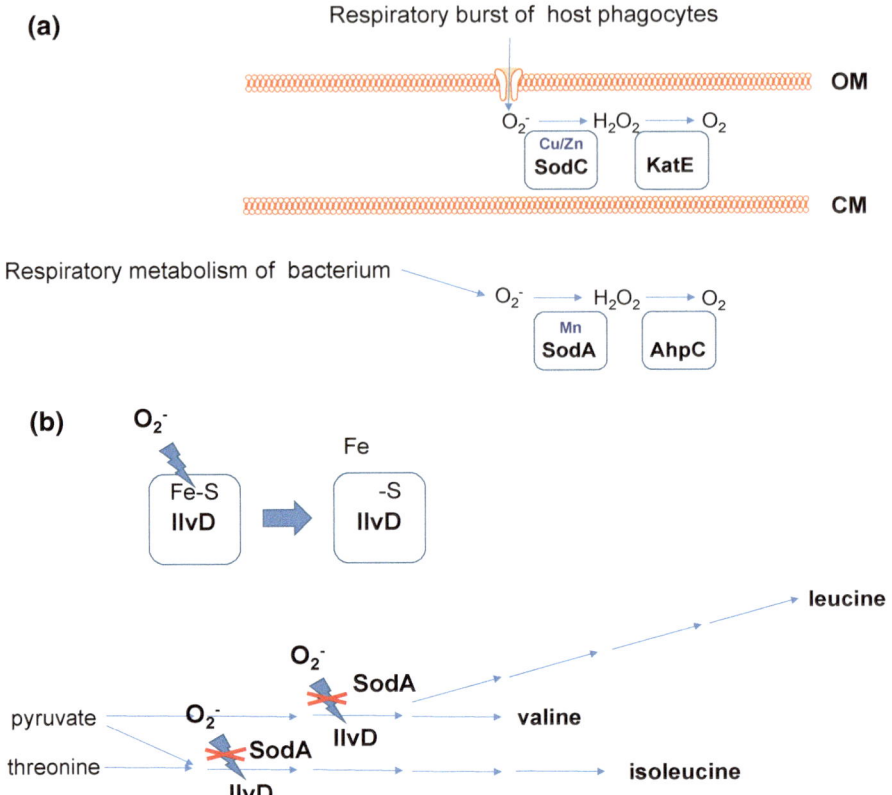

Fig. 3.5 **a** Proposed roles of SodC and SodA in protecting *Brucella* strains from superoxide of exogenous and endogenous origin, and **b** and proposed role of SodA in protecting the branched chain amino acid biosynthesis pathway in bacteria. KatE—periplasmic catalase; AhpC—periplasmic peroxiredoxin; IlvD—dihydroxyisovalerate dehydratase

SodA than its parent when these strains are cultivated in a Mn-deprived medium (Anderson et al. 2009), and the *mntH* mutant is also considerably more sensitive than the parent to the endogenous O_2^- generators menadione and paraquat. One of the major functions of the *Brucella* SodA appears to be to protect amino acid biosynthetic enzymes from inactivation by endogenous O_2^- generated by this bacterium's respiratory metabolism (Martin et al. 2012). The Fe-S centers in the branched chain amino acid biosynthesis enzyme dihydroxyisovalerate dehydratase (IlvD) are well-documented targets of O_2^- inactivation (Kuo et al. 1987; Brown et al. 1995), and protecting these enzymes is an established function of bacterial cytoplasmic SODs (Carlioz and Touati 1986; Brown et al. 1995; Imlay 2003) (Fig. 3.5b). The fact that a *B. abortus sodA* mutant displays auxotrophy for amino acids in addition to isoleucine, leucine and valine, however, suggests that metabolic

Fig. 3.6 Spleen colonization profiles of *B. abortus* 2308 and an isogenic *sodA* mutant in C57BL/6[Nramp+/+] mice. Mice were infected with 5×10^4 brucellae via the intraperitoneal route

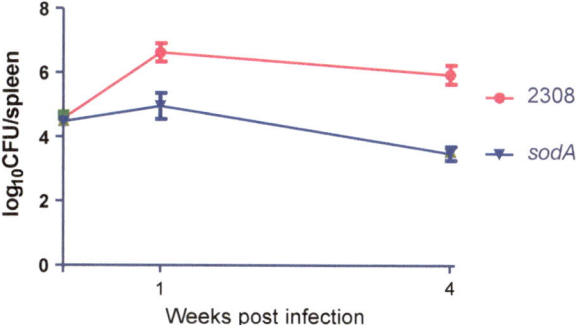

enzymes in addition to IlvD are also prone to O_2^- inactivation in *Brucella* (Martin et al. 2012).

SodA is required for the wild-type virulence of *B. abortus* 2308 in mice (Fig. 3.6) (Martin et al. 2012). As noted previously, the cytoplasmic location of this antioxidant theoretically precludes it from providing a direct defense against O_2^- generated by host phagocytes, and its inability to do so has been confirmed experimentally (Martin et al. 2012). Instead, it seems likely that SodA's capacity to protect IlvD and other enzymes involved in amino acid biosynthesis from endogenous O_2^- underlies its role in virulence. This proposed function fits with the fact that *ilvD* and other amino acid biosynthesis genes have been shown to be required for the virulence of *Brucella* strains (Foulongne et al. 2000; Lestrate et al. 2000, 2003; Köhler et al. 2002; Alcantara et al. 2004).

3.3.2 PykM

Like its *B. abortus* counterpart, a *Bradyrhizobium japonicum mntH* mutant exhibits a significant growth defect during routine aerobic cultivation that can be rescued by supplementing the growth medium with Mn (Hohle and O'Brian 2012). One important contributor to this phenotype is PykM, the sole pyruvate kinase produced by *B. japonicum*, which is a Mn-dependent enzyme. PykM catalyzes the conversion of phosphoenolpyruvate to pyruvate in the lower portion of the glycolytic pathway, and reduced pyruvate kinase activity limits the ability of the *B. japonicum mntH* mutant to catabolize certain C and E sources such as glycerol.

Brucella strains have a blocked glycolytic pathway because they lack the enzyme phosphofructokinase (Robertson and McCullough 1968a, b). Biochemical and isotopic labeling studies also indicate that most of the brucellae have an inoperative Entner-Duodoroff pathway (Robertson and McCullough 1968a, b; T. Barbier, personal communication), although genomic analyses indicate that they possess the

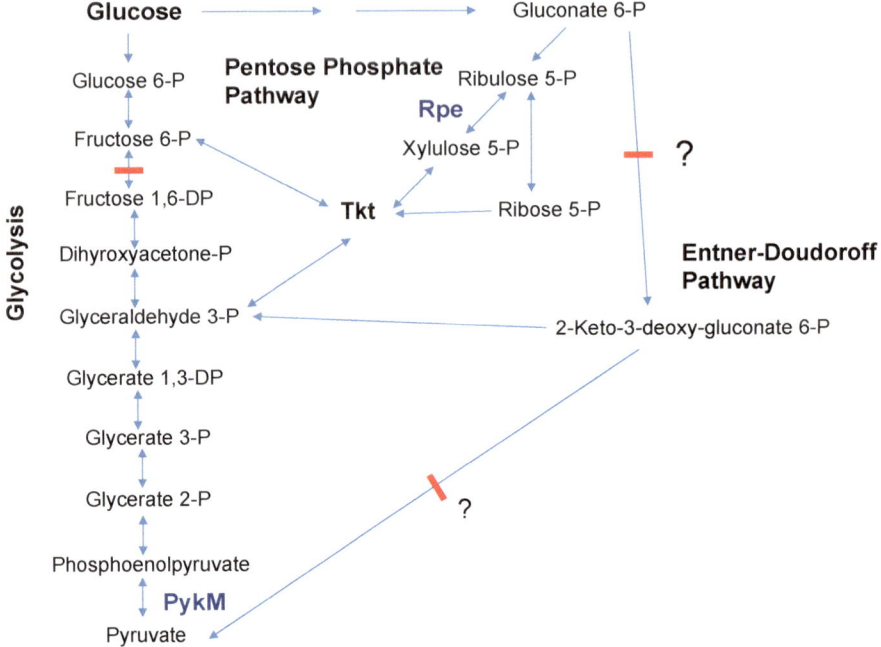

Fig. 3.7 A blocked glycolytic and inoperative Entner-Duordoroff pathways force most *Brucella* strains to rely upon the pentose phosphate pathway for the catabolism of glucose. The '?' next to the blocks in the pathways denoted by the *red bars* indicates that is currently unknown why the ED pathway is inoperative in these bacteria since genes required for its operation are present. The proposed roles of Mn^{2+} in the activities of pyruvate kinase (PykM) and ribulose-5-phosphate 3-epimerase (Rpe) is discussed in the text

genes necessary to encode the enzymes required to use this alternative pathway for hexose catabolism. The absence of both of these pathways theoretically forces these bacteria to rely exclusively on the pentose phosphate pathway for the catabolism of glucose (Essenberg et al. 2002), which would require the activity of pyruvate kinase as depicted in Fig. 3.7. Notably, the single pyruvate kinase produced by *Brucella* strains is a PykM homolog (Pitzer and Baumgartner, unpublished data), and *B. abortus pykM* mutants display significant attenuation in both BALB/c (Gao et al. 2016) and C57BL/6 mice (Fig. 3.8). The proposed link between PykM activity and Mn is therefore an important consideration because glucose catabolism has been shown to be required for the virulence of *Brucella* strains in mice (Xavier et al. 2013). Specifically, it is conceivable that one of the important roles that MntH plays in the virulence of *Brucella* is to provide these bacteria with sufficient levels of Mn to support PykM activity, and allow the efficient catabolism of glucose.

Fig. 3.8 Spleen colonization profiles of *B. abortus* 2308 and an isogenic *pykM* mutant in C57BL/6$^{Nramp+/+}$ mice. Mice were infected with 5×10^4 brucellae via the intraperitoneal route

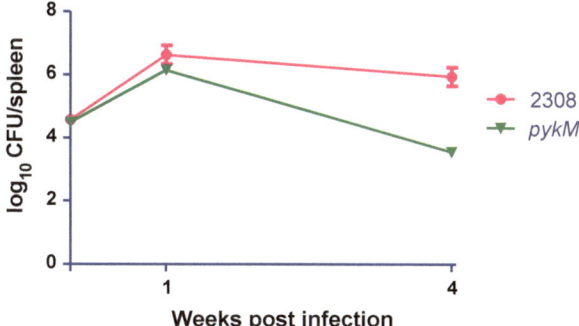

3.3.3 Rsh

Many bacteria elicit a physiologic response known as the 'stringent response' when they are faced with nutrient deprivation. In response to a variety of endogenous environmental stimuli resulting from different types of 'starvation', bacterial cells begin producing the signaling molecules guanosine 3'-diphosphate 5'-diphosphate (ppGpp) and guanosine 3'-diphosphate 5'-triphosphate (pppGpp), collectively known as (p)ppGpp (Boutte and Crosson 2013; Hauryliuk et al. 2015). When these so-called 'alarmones' bind to the β-subunit of RNA polymerase it changes the specificity of RNAP for promoters (Barker et al. 2001a, b) causing a programmatic shift in gene expression that switches the cell's physiology from one geared toward growth and replication, to one directed toward stress resistance and maintenance of cell viability (Magnusson et al. 2005). In some bacteria, the stringent response also induces the expression of genes encoding adhesins, toxins, secretion systems and other factors required for the successful symbiotic or pathogenic interactions of these bacteria with their eukaryotic hosts (Wells and Long 2002; Dalebroux et al. 2010).

The (p)ppGpp synthetase RelA and the bifunctional (p)ppGpp synthetase/ hydrolase SpoT control the stringent response in *Escherichia coli* and a limited number of other bacteria. But most bacteria, including *Brucella* (Dozot et al. 2006), rely upon a single bifunctional SpoT homolog known as Rel or Rsh (RelA-SpoT hybrid) to regulate their stringent response (Atkinson et al. 2011; Hauryliuk et al. 2015), and the hydrolase activity of this enzyme is Mn-dependent (De Boer et al. 1977; Heinemeyer et al. 1978; Johnson et al. 1979; Mechold et al. 1996). *Brucella rsh* mutants are highly attenuated in mice and cultured mammalian cells (Kim et al. 2005; Dozot et al. 2006), and the basis for this phenotype appears to be two-fold. First, the growth defects displayed by these mutants in vitro suggests that they are unlikely to be able to withstand the nutrient deprivation they encounter in their intracellular niche in host macrophages (Köhler et al. 2003; Roop et al. 2009). Second, Rsh is also required for the proper expression of the genes encoding the Type IV secretion system (T4SS) in *Brucella* strains (Dozot et al. 2006; Hanna et al. 2013), and the T4SS secretes effectors into the host cell cytoplasm that re-direct the intracellular trafficking of the *Brucella* containing vacuoles (BCVs) and facilitate the formation of the replicative BCVs within which the brucellae maintain their intracellular persistence (de Jong and Tsolis 2012; Lacerda et al. 2013; Celli 2015). Thus, it is quite possible that another way that MntH contributes to virulence is by providing sufficient levels of Mn allow the *Brucella* Rsh to properly balance its (p) pGGpp synthetase and hydrolase activities during induction and resolution of the 'stringent response'.

3.3.4 Pcs

Phosphatidylcholine (PC) is a major component of eukaryotic cell membranes, but is not a major component of most bacterial membranes (Sohlenkamp et al. 2003). The α-proteobacteria are an exception however, where PC in the cell envelope is thought to be important for their interactions with their eukaryotic hosts (Aktas et al. 2010). *Agrobacterium tumefaciens* and *B. abortus* mutants that do not produce PC, for instance, are attenuated in plants (Wessel et al. 2006) and mice (Comerci et al. 2006; Conde-Alvarez et al. 2006), respectively, and *Bradyrhizobium japonicum* PC-deficient mutants do not form wild-type symbiotic nodules on soybeans (Minder et al. 2001). The *Agrobacterium* phosphatidycholine synthase (Pcs) is a Mn-dependent enzyme (Aktas et al. 2014), and if the same holds true for its *Brucella* counterpart, its role in PC biosynthesis may be another reason that Mn is such an important micronutrient for these bacteria in both the in vitro and in vivo settings.

3.3.5 BpdA and BpdB

Cyclic-di-GMP (c-di-GMP) is an endogenous signaling molecule that globally regulates both gene expression and gene product function in bacteria (Römling et al. 2013), and c-di-GMP regulated genes and gene products are important virulence determinants for many bacterial pathogens (Ryan 2013). Cellular c-di-GMP levels are controlled by the activities of c-di-GMP synthases (also known as diguanylate cyclases) and c-di-GMP-specific phosphodiesterases. Petersen et al. (2011) showed that two c-di-GMP-specific phosphodiesterases, BpdA and BpdB, are required for the wild-type virulence of *B. melitensis* 16 M in immunocompromised mice, and that BpdA plays a role in proper expression of the flagellar biosynthesis genes in this strain. Interestingly, the c-di-GMP phosphodiesterases VieA from *Vibrio cholerae* (Tamayo et al. 2005) and HmsP from *Yersinia pestis* (Bobrov et al. 2005) are both Mn-dependent enzymes, and it will be important to determine if the same holds true for BpdA and BpdB, and if MntH plays a role in supporting the activities of these proteins.

3.4 Manganese as an Antioxidant

In addition to its serving as a co-factor for multiple enzymes, Mn has also been shown to be an important antioxidant in bacteria. Mn^{2+} in complex with small metabolic anions such as phosphate and lactate, for instance, can directly catalyze the detoxification of O_2^- (Culotta and Daly 2013), and such complexes play an important role in oxidative defense in bacteria such as *Lactobacillus plantarum* (Archibald and Fridovich 1981) and *Deinococcus radiodurans* (Slade and Radman 2011) that are capable of accumulating high (e.g. mM) concentrations of intracellular Mn. Recent studies with *E. coli*, a bacterium that does not rely heavily on Mn for its basic metabolism, have also shown that Mn can serve as an indirect defense against oxidative damage (Anjem et al. 2009). This bacterium increases its Mn import in response to exposure to ROS, and substitutes this metal for Fe in enzymes and proteins with mononuclear Fe centers (Sobota and Imlay 2011). Unlike Fe, Mn does not actively catalyze the production of OH^- from O_2^- and H_2O_2 under physiologic conditions. Consequently, this substitution protects these proteins from oxidative damage due to Fenton chemistry. These Mn-based ROS defenses explain why some bacteria increase the expression of their Mn transport genes in response to oxidative stress, and these genes are often controlled by H_2O_2-responsive transcriptional regulators such as OxyR and PerR (Kehres et al. 2002a, b; Anjem et al. 2009; Wang et al. 2014).

Experimental evidence suggests that *Brucella* strains may also employ a Mn for Fe substitution defense to protect key cellular proteins from oxidative damage. Expression of the gene encoding the Mn transporter MntH, for instance, is elevated in *B. abortus* 2308 in response to exposure to both exogenous and endogenous sources of H_2O_2 (Fig. 3.9), and a *B. abortus mntH* mutant exhibits increased sensitivity to this ROS compared to its parent strain (Anderson et al. 2009). One notable candidate for this type of protection is the enzyme ribulose-5-phosphate 3-epimerase (Rpe), which catalyzes a critical step in the pentose phosphate pathway (Fig. 3.7). This enzyme is one of the key targets for this type of ROS defense in *E. coli* (Sobota and Imlay 2011), and as noted in a previous section, this pathway is thought to be essential for hexose catabolism in *Brucella* strains. Moreover, attempts to construct *rpe* mutants in two different laboratories have led to the conclusion that this gene is essential (Pitzer, unpublished data; De Bolle, personal communication). Thus, another conceivable role that MntH might be playing in *Brucella* is to provide increased intracellular levels of Mn as an indirect defense against the oxidative damage to cellular proteins.

3.5 Manganese Homeostasis

Mn occupies a relatively low position on the Irwin-Williams scale of reactivity compared to other biologically relevant metals (Waldron and Robinson 2009). Because Mn does not catalyze Fenton chemistry under physiologic conditions, and some bacteria tolerate high intracellular concentrations of Mn without apparent signs of toxicity, until recently Mn was considered by many to be of relatively low toxicity for prokaryotes. However, high affinity Mn import is tightly regulated in bacteria by Mn-responsive regulators such as MntR (Que and Helmann 2000; Patzer and Hantke 2001) and Mur (Díaz-Mireles et al. 2005), and prokaryotic Mn exporters have also been described (Rosch et al. 2009; Li et al. 2011; Veyrier et al. 2011; Waters et al. 2011; Cubillas et al. 2014; Raimunda and Elso-Berberián 2014; Martin et al. 2015). Moreover, the phenotypes displayed by bacterial mutants that lack Mn-responsive regulators or Mn exporters (Kitphati et al. 2007) clearly show that that excess Mn intracellular Mn is indeed toxic.

Mn, Mg and Fe are interchangeable in the active sites of many bacterial proteins (Jakubovics and Jenkinson 2001), and accordingly, the targets of Mn toxicity that have been identified in prokaryotes tend to be proteins involved in Mg or Fe transport, or proteins that require one of these metals for their activity. Excess levels of Mn, for instance, can interfere with Mg transport in *Escherichia coli* (Silver et al. 1972) and *Bradyrhizobium japonicum* (Hohle and O'Brian 2014), and an imbalanced intracellular ratio of Mn to Mg can also inhibit the activity of the TCA cycle enzyme isocitrate lyase in *B. japonicum*. High levels of intracellular Mn compared to Fe have also been shown to prevent the Fe-responsive regulator Fur from being able to properly recognize intracellular Fe levels in *E. coli* (Hantke 1987; Martin et al. 2015), and interfere with the ability of the H_2O_2-responsive regulator PerR to recognize ROS stress in *Bacillus subtilis* (Herbig and Helmann 2001) and *Streptococcus pyogenes* (Turner et al. 2015).

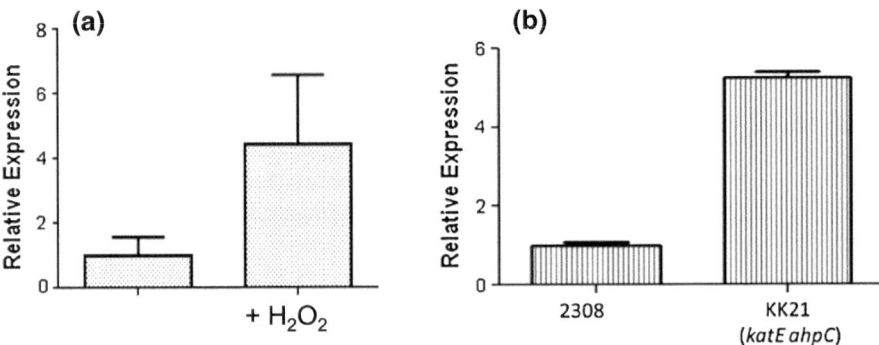

Fig. 3.9 **a** Induction of *mntH* transcription in *B. abortus* 2308 following exposure to 5 mM H_2O_2; and **b** relative levels of *mntH* transcription in *B. abortus* 2308 and an isogenic *katE ahpC* mutant [which produces 'elevated' levels of endogenous H_2O_2 compared to 2308 during routine in vitro cultivation (Steele et al. 2010)]

3.5.1 *Mur*

The gene encoding the *Brucella* manganese uptake regulator (Mur) was originally identified based on its capacity to restore the ability of an *E. coli fur* mutant to regulate an Fe-responsive gene fusion (Phillips et al. 1996). Subsequent studies, however, showed that gene does not directly regulate *Brucella* Fe acquisition genes (Freeland et al. 1996; Bellaire 2001). Rather, like its counterpart in *R. legumi-nosarum* (Díaz-Mireles et al. 2004) and several other α–proteobacteria (Chao et al. 2004; Platero et al. 2004; Rodionov et al. 2006; Kitphati et al. 2007; Hohle and O'Brian 2009; Heindl et al. 2016), this protein functions as a Mn-responsive regulator as its name implies. To date, the only Mur-repressed gene that has been identified in *Brucella* is *mntH* (Anderson et al. 2009), where Mur binds specifically to a 28 nt region in the *mntH* promoter and represses the expression of this gene in response to increasing levels of Mn (Menscher et al. 2012) (Fig. 3.10). Interestingly, nucleotide sequences displaying more than limited consensus with the Mur binding site appear to be rare in the *B. abortus* 2308 genome sequence, suggesting that the Mur regulon in *Brucella* may be small, and perhaps, limited to *mntH* and the gene designated *irr1* by Rodionov et al. (2006) and currently annotated as BAB_RS17810 in the *B. abortus* 2308 genome sequence. It is important to note that this gene does not encode the *Brucella* Irr (Martínez et al. 2005), but is predicted to encode another transcriptional regulator belonging to the

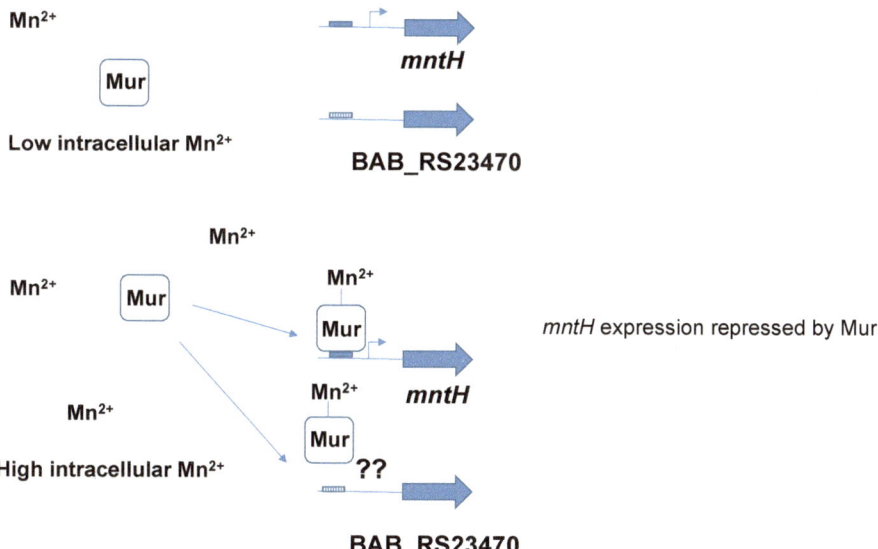

Fig. 3.10 The transcriptional repressor Mur represses mntH expression in *Brucella* strains in a Mn-responsive fashion. The *filled box* in the *mntH* promoter region indicates the Mur binding site, and the *hatched box* upstream of the BAB_RS23470 gene denotes the presence of a 'predicted' Mur binding site that has not been experimentally confirmed

Fur family whose function has not yet been determined (Menscher et al. 2012). Given the relationship between the Fur homolog PerR and Mn homeostasis in other bacteria (Helmann 2014; Wang et al. 2014), and fact that *mntH* expression in *B. abortus* is H_2O_2-responsive, one intriguing possibility is that this gene encodes a *Brucella* PerR homolog.

3.5.2 EmfA

The observation that a *B. abortus mur* mutant exhibits wild-type resistance to high levels of Mn in the growth medium (Pitzer, unpublished data) suggested that a Mn exporter might be present in *Brucella*. Indeed, a search of the *Brucella* genome sequences currently available in GenBank identified a gene (designated BAB_RS23470) predicted to encode a homolog of the *Rhizobium etli* Mn exporter EmfA (Cubillas et al. 2014), which belongs to the cation diffusion family of metal exporters. Phenotypic evaluation of a *B. abortus emfA* mutant is currently underway to determine if the corresponding gene product is in fact a Mn^{2+} exporter.

3.6 Conclusions

High affinity Mn transport is essential for the wild-type virulence of *Brucella* strains, but the reason that this metal plays such an important role in the basic biology and pathogenesis of these bacteria has only been partially explained. Further studies on Mn homeostasis are needed to help us better understand this relationship.

Acknowledgements Work on Mn metabolism in *Brucella* in the laboratory of RMR has been funded by grants AI48499, AI63516 and AI112745 from the National Institute of Allergy and Infectious Disease and internal funding from the Office of Research and Graduate Studies, Brody School of Medicine, East Carolina University.

References

Aktas M, Wessel M, Hacker S, Klüsener S, Gleichenhagen J, Narberhaus F (2010) Phosphatidylcholine biosynthesis and its significance in bacteria interacting with eukaryotic cells. Eur J Cell Biol 89:888–894

Aktas M, Köster S, Kizilirmak S, Casanova JC, Betz H, Fritz C, Moser R, Yildiz O, Narberhaus F (2014) Enzymatic properties and substrate specificity of a bacterial phosphatidylcholine synthase. FEBS J 281:3523–3541

Alcantara RB, Read RDA, Valderas MW, Brown TD, Roop RM II (2004) Intact purine biosynthesis pathways are required for wild-type virulence of *Brucella abortus* 2308 in the BALB/c mouse model. Infect Immun 72:4911–4917

Anderson ES, Paulley JT, Gaines JM, Valderas MW, Martin DW, Menscher E, Brown TD, Burns CS, Roop RM II (2009) The manganese transporter MntH is a critical virulence determinant for *Brucella abortus* 2308 in experimentally infected mice. Infect Immun 77:3466–3474

Anjem A, Varghese S, Imlay JA (2009) Manganese import is a key element of the OxyR response to hydrogen peroxide in *Escherichia coli*. Mol Microbiol 72:844–858

Archibald FS, Duong MN (1984) Manganese acquisition by *Lactobacillus plantarum*. J Bacteriol 158:1–8

Archibald FS, Fridovich I (1981) Manganese and defenses against oxygen toxicity in *Lactobacillus plantarum*. J Bacteriol 145:442–451

Arpaia N, Godec J, Lau L, Sivick KE, McLaughlin LM, Jones MB, Dracheva T, Peterson SN, Monack DM, Barton GM (2011) TLR signaling is required for *Salmonella typhimurium* virulence. Cell 144:675–688

Atkinson GC, Tenson T, Hauryliuk V (2011) The RelA/SpoT homolog (RSH) superfamily: distribution and functional evolution of ppGpp synthetases and hydrolases across the tree of life. PLoS ONE 6:e23479

Barker MM, Gaal T, Josaitis CA, Gourse RL (2001a) Mechanism of regulation of transcription initiation by ppGpp. I. Effects of ppGpp on transcription in vivo and in vitro. J Mol Biol 305:673–688

Barker MM, Gaal T, Josaitis CA, Gourse RL (2001b) Mechanism of regulation of transcription initiation by ppGpp. II. Models for positive control based on properties of RNAP mutants and competition for RNAP. J Mol Biol 305:689–702

Bearden SW, Perry RD (1999) The Yfe system of *Yersinia pestis* transports iron and manganese and is required for full virulence of plague. Mol Microbiol 32:403–414

Beck BL, Tabatabai LB, Mayfield JE (1990) A protein isolated from *Brucella abortus* is a Cu-Zn superoxide dismutase. Biochemistry 29:372–376

Bellaire BH (2001) Production of the siderophore 2,3-dihydroxybenzoic acid by *Brucella abortus* is regulated independent of Fur and is required for virulence in cattle. Doctoral dissertation, Louisiana State University Health Sciences Center—Shreveport

Berntsson RPA, Smits SHJ, Schmitt L, Slotboom D-J, Poolman B (2010) A structural classification of substrate-binding proteins. FEBS Lett 884:2606–2617

Bobrov AG, Kirillina O, Perry RD (2005) The phosphodiesterase activity of the HmsP EAL domain is required for negative regulation of biofilm formation in *Yersinia pestis*. FEMS Microbiol Lett 247:123–130

Boutte CC, Crosson S (2013) Bacterial lifestyle shapes stringent response activation. Trends Microbiol 21:174–180

Bricker BJ, Tabatabai LB, Judge BA, Deyoe BL, Mayfield JE (1990) Cloning, expression, and occurrence of the *Brucella* Cu-Zn superoxide dismutase. Infect Immun 58:2935–2939

Brown OR, Smyk-Randall E, Draczynska-Lusiak B, Fee JA (1995) Dihydroxy-acid dehydratase, a [4Fe-4S] cluster-containing enzyme in *Escherichia coli*: effects of intracellular superoxide dismutase on its inactivation by oxidant stress. Arch Biochem Biophys 319:10–22

Carlioz A, Touati D (1986) Isolation of superoxide dismutase mutants in *Escherichia coli*: is superoxide dismutase necessary for aerobic life? EMBO J 5:623–630

Celli J (2015) The changing nature of the *Brucella*-containing vacuole. Cell Microbiol 17:951–958

Cellier M, Govoni G, Vidal S, Kwan T, Groulx N, Liu J, Sanchez F, Skamene E, Schurr E, Gros P (1994) Human natural resistance-associated macrophage protein: cDNA cloning, chromosomal mapping, genomic organization, and tissue-specific expression. J Exp Med 180:1741–1752

Cellier MF, Courville P, Campion C (2007) Nramp1 phagocyte intracellular metal withdrawal defense. Microbes Infect 9:1662–1670

Champion OL, Karlyshev A, Cooper IAM, Ford DC, Wren BW, Duffield M, Oyston PCF, Titball RW (2011) *Yersinia pseudotuberculosis mntH* functions in intracellular manganese accumulation, which is essential for virulence and survival in cells expressing functional Nramp1. Microbiology 157:1115–1122

Chao TC, Becker A, Buhrmeister J, Pühler A, Weidner S (2004) The *Sinorhizobium meliloti fur* gene regulates, with dependence on Mn(II), transcription of the *sitABCD* operon, encoding a metal-type transporter. J Bacteriol 186:3609–3620

Comerci DJ, Altabe S, de Mendoza D, Ugalde R (2006) *Brucella abortus* synthesizes phosphatidylcholine from choline provided by the host. J Bacteriol 188:1929–1934

Conde-Alvarez R, Grilló MJ, Salcedo SP, de Miguel MJ, Fugier E, Gorvel JP, Moriyón I, Iriarte M (2006) Synthesis of phosphatidylcholine, a typical eukaryotic phospholipid, is necessary for full virulence of the intracellular bacterial parasite *Brucella abortus*. Cell Microbiol 8:1322–1335

Corbin BD, Seeley EH, Raab A, Feldmann J, Miller MR, Torres VJ, Anderson KL, Dattilo M, Dunman PM, Gerads R, Caprioli RM, Nacken W, Chazin WJ, Skaar EP (2008) Metal chelation and inhibition of bacterial growth in tissue abscesses. Science 319:962–965

Cotruvo JA, Stubbe J (2010) An active dimanganese(III)-tyrosyl radical cofactor in *Escherichia coli* class Ib ribonucleotide reductase. Biochemistry 49:1297–1309

Crump KE, Bainbridge B, Brusko S, Turner LS, Ge X, Stone V, Xu P, Kitten T (2014) The relationship of the lipoprotein SsaB, manganese, and superoxide dismutase in *Streptococcus sanguinis* virulence for endocarditis. Mol Microbiol 92:1243–1259

Cubillas C, Vinuesa P, Tabche ML, Dávalos A, Vázquez A, Hernández-Lucas I, Romero D, García-de los Santos A (2014) The cation diffusion facilitator protein EmfA of *Rhizobium etli* belongs to a novel subfamily of Mn^{2+}/Fe^{2+} transporters conserved in α-proteobacteria. Metallomics 6:1808–1815

Culotta VC, Daly MJ (2013) Manganese complexes: diverse metabolic routes to oxidative stress resistance in prokaryotes and yeast. Antioxid Redox Signal 19:933–945

Dalebroux ZD, Svensson SL, Gaynor EC, Swanson MS (2010) ppGpp conjures bacterial virulence. Microbiol Mol Biol Rev 74:171–199

Davies BW, Walker GC (2007) Disruption of *sitA* compromises *Sinorhizobium meliloti* for manganese uptake required for protection against oxidative stress. J Bacteriol 189:2101–2109

De Boer HA, Bakker AJ, Gruber M (1977) Breakdown of ppGpp in $spoT^+$ and $spoT^-$ cells of *Escherichia coli*. Manganese and energy requirement and tetracycline inhibition. FEBS Lett 79:19–24

de Jong MF, Tsolis RM (2012) Brucellosis and type IV secretion. Future Med 7:47–58

Díaz-Mireles E, Wexler M, Sawers G, Bellini D, Todd JD, Johnston AWB (2004) The Fur-like protein Mur of *Rhizobium leguminosarum* is a Mn^{2+}-responsive transcriptional regulator. Microbiology 150:1447–1456

Dozot M, Boigegrain RA, Delrue RM, Hallez R, Ouahrani-Bettache S, Danese I, Letesson JJ, De Bolle X, Köhler S (2006) The stringent response mediator Rsh is required for *Brucella melitensis* and *Brucella suis* virulence, and for expression of the type IV secretion system virB. Cell Microbiol 8:1791–1802

Essenberg RC, Seshadri R, Nelson K, Paulsen I (2002) Sugar metabolism by Brucellae. Vet Microbiol 90:249–261

Evenson MA, Gerhardt P (1955) Nutrition of brucellae: utilization of iron, magnesium and manganese for growth. Proc Soc Exp Biol Med 89:678–680

Foulongne V, Bourg G, Cazevieille G, Michaux-Charachon S, O'Callaghan D (2000) Identification of *Brucella suis* genes affecting intracellular survival in an in vitro human macrophage infection model by signature-tagged transposon mutagenesis. Infect Immun 68:1297–1303

Freeland RL, Kovach ME, Roop RM II, Baldwin CL, Elzer PH (1996) Molecular characterization of the *Brucella abortus* ferric uptake regulator (*fur*) locus. In: Proceedings of 77th Annual Conference of Research Workers in Animal Disease, Abstract 4, p 6

Gao K, Tian M, Bao Y, Li P, Liu J, Ding C, Wang S, Li T, Yu S (2016) Pyruvate kinase is necessary for *Brucella abortus* full virulence in BALB/c mouse. Vet Res 47:e87

Gee JM, Valderas MW, Kovach ME, Grippe VK, Robertson GT, Ng W-L, Richardson JM, Winkler ME, Roop RM II (2005) The *Brucella abortus* Cu, Zn superoxide dismutase is required for optimal resistance to oxidative killing by murine macrophages and wild-type virulence in experimentally infected mice. Infect Immun 73:2873–2880

Gunshin H, Mackenzie B, Berger UV, Gunshin Y, Romero MF, Boron WF, Nussberger S, Gollan JL, Hediger MA (1997) Cloning and characterization of a mammalian proton-coupled metal-ion transporter. Nature 388:482–488

Hanna N, Ouahrani-Bettache S, Drake KL, Adams LG, Köhler S, Occhialini A (2013) Global Rsh-dependent transcription profile of *Brucella suis* during stringent response unravels adaptation to nutrient starvation and cross-talk with other stress responses. BMC Genom 14: e459

Hantke K (1987) Selection procedure for deregulated iron transport mutants (*fur*) in *Escherichia coli* K12: *fur* not only affects iron metabolism. Mol Gen Genet 210:135–139

Hassan HM, Fridovich I (1979) Paraquat and *Escherichia coli*—mechanism of production of extracellular superoxide radical. J Biol Chem 254:10846–10852

Hauryliuk V, Atkinson GC, Murakami KS, Tenson T, Gerdes K (2015) Recent functional insights into the role of (p)ppGpp in bacterial physiology. Nat Rev Microbiol 13:298–309

Heindl JE, Hibbing ME, Xu J, Natarajan R, Buechlein AM, Fuqua C (2016) Discrete responses to limitation for iron and manganese in *Agrobacterium tumefaciens*: influence on attachment and biofilm formation. J Bacteriol 198:816–829

Heinemeyer EA, Geis M, Richter D (1978) Degradation of guanosine 3′-diphosphate 5′-diphosphate in vitro by the *spoT* gene product of *Escherichia coli*. Eur J Biochem 89:125–131

Helmann JD (2014) Specificity of metal sensing: iron and manganese homeostasis in *Bacillus subtilis*. J Biol Chem 289:28112–28120

Herbig AF, Helmann JD (2001) Roles of metal ions and hydrogen peroxide in modulating the interaction of the *Bacillus subtilis* PerR peroxide regulon repressor with operator DNA. Mol Microbiol 41:849–859

Hohle TH, O'Brian MR (2009) The *mntH* gene encodes the major Mn^{2+} transporter in *Bradyrhizobium japonicum* and is regulated by manganese via the Fur protein. Mol Microbiol 72:399–409

Hohle TH, O'Brian MR (2012) Manganese is required for oxidative metabolism in unstressed *Bradyrhizobium japonicum* cells. Mol Microbiol 84:766–777

Hohle TH, O'Brian MR (2014) Magnesium-dependent processes are targets of bacterial manganese toxicity. Mol Microbiol 93:736–747

Horsburgh MJ, Wharton SJ, Cox AG, Ingham E, Peacock S, Foster SJ (2002) MntR modulates expression of the PerR regulon and superoxide resistance in *Staphylococcus aureus* through control of manganese uptake. Mol Microbiol 44:1269–1286

Imlay J (2003) Pathways of oxidative damage. Annu Rev Microbiol 57:395–418

Imlay J (2008) Cellular defenses against superoxide and hydrogen peroxide. Annu Rev Biochem 77:755–776

Jakubovics NS, Jenkinson HF (2001) Out of the iron age: new insights into the critical role of manganese homeostasis in bacteria. Microbiology 147:1709–1718

Janulczyk R, Ricci S, Björck L (2003) MtsABC is important for manganese and iron transport, oxidative stress resistance, and virulence of *Streptococcus pyogenes*. Infect Immun 71:2656–2664

Johnson GS, Adler CR, Collins JJ, Court D (1979) Role of the *spoT* gene product and manganese ion in the metabolism of guanosine 5′-diphosphate 3′-diphosphate in *Escherichia coli*. J Biol Chem 254:5483–5487

Kehl-Fie TE, Skaar EP (2012) Nutritional immunity beyond iron: a role for manganese and zinc. Curr Opin Chem Biol 14:218–224

Kehl-Fie TE, Chitayat S, Hood MI, Damo S, Restrepo N, Garcia C, Munro KA, Chazin WJ, Skaar EP (2011) Nutrient metal sequestration by calprotectin inhibits bacterial superoxide defense, enhancing neutrophil killing of *Staphylococcus aureus*. Cell Host Microbe 10: 158–164

Kehl-Fie TE, Zhang Y, Moore JL, Farrand AJ, Hood MI, Rathi S, Chazin WJ, Caprioli RM, Skaar EP (2013) MntABC and MntH contribute to systemic *Staphylococcus aureus* infection by competing with calprotectin for nutrient manganese. Infect Immun 81:3395–3405

Kehres DG, Zaharik ML, Finlay BB, Maguire ME (2000) The NRAMP proteins of *Salmonella typhimurium* and *Escherichia coli* are selective manganese transporters involved in the response to reactive oxygen. Mol Microbiol 36:1085–1100

Kehres DG, Janakiraman A, Slauch JM, Maguire ME (2002a) Regulation of *Salmonella enterica* serovar Typhimurium *mntH* transcription by H_2O_2, Fe^{2+}, and Mn^{2+}. J Bacteriol 184: 3151–3158

Kehres DG, Janakiraman A, Slauch JM, Maguire ME (2002b) SitABCD is the alkaline Mn^{2+} transporter of *Salmonella enterica* serovar Typhimurium. J Bacteriol 184:3159–3166

Kim S, Watanabe K, Suzuki H, Watarai M (2005) Roles of *Brucella abortus* SpoT in morphological differentiation and intramacrophagic replication. Microbiology 151:1607–1617

Kitphati W, Ngok-ngam P, Suwanmaneerat S, Sukchawalit R, Mongkolsuk S (2007) *Agrobacterium tumefaciens fur* has important physiologic roles in iron and manganese homeostasis, the oxidative stress response, and full virulence. Appl Environ Microbiol 73:4760–4768

Köhler S, Foulongne V, Ouahrani-Bettache S, Bourg G, Teyssier J, Ramuz M, Liautard JP (2002) The analysis of the intramacrophagic virulome of *Brucella suis* deciphers the environment encountered by the pathogen inside the macrophage host cell. Proc Natl Acad Sci U S A 99:15711–15716

Köhler S, Michaux-Charachon S, Porte F, Ramuz M, Liautard JP (2003) What is the nature of the replicative niche of a stealthy bug named *Brucella*? Trends Microbiol 11:215–219

Kuo CF, Mashino T, Fridovich I (1987) α,β-dihydroxyisovalerate dehydratase: a superoxide-sensitive enzyme. J Biol Chem 262:4724–4727

Lacerda TLS, Salcedo SP, Gorvel JP (2013) *Brucella* T4SS: the VIP pass inside host cells. Curr Opin Microbiol 16:45–51

Lee SA, Gallagher LA, Thongdee M, Staudinger BJ, Lippman S, Singh PK, Manoil C (2015) General and condition-specific essential functions of *Pseudomonas aeruginosa*. Proc Natl Acad Sci U S A 112:5189–5194

Lessie TG, Phibbs PV (1984) Alternative pathways of carbohydrate utilization in pseudomonads. Ann Rev Microbiol 38:359–387

Lestrate P, Delrue RM, Danese I, Didembourg C, Taminiau B, Mertens P, De Bolle X, Tibor A, Tang CM, Letesson JJ (2000) Identification and characterization of in vivo attenuated mutants of *Brucella melitensis*. Mol Microbiol 38:543–551

Lestrate P, Dricot A, Delrue RM, Lambert C, Martinelli V, De Bolle X, Letesson JJ, Tibor A (2003) Attenuated signature-tagged mutagenesis mutants of *Brucella melitensis* identified during the acute stage of infection in mice. Infect Immun 71:7053–7060

Li C, Tao J, Mao D, He C (2011) A novel manganese efflux system, YebN, is required for virulence by *Xanthomonas oryzae* pv. *oryzae*. PLoS ONE 6:e21983

Lisher JP, Giedroc DP (2013) Manganese acquisition and homeostasis at the host-pathogen interface. Front Cell Infect Microbiol 3:e91

Magnusson LU, Farewell A, Nystrom T (2005) ppGpp: a global regulator in *Escherichia coli*. Trends Microbiol 13:236–242

Malo D, Vogan K, Vidal S, Hu J, Cellier M, Schurr E, Fuks A, Bumstead N, Morgan K, Gros P (1994) Haplotype mapping and sequence analysis of the mouse *Nramp* gene predict susceptibility to infection with intracellular parasites. Genomics 23:51–61

Makui H, Roig E, Cole ST, Helmann JD, Gros P, Cellier MFM (2000) Identification of the *Escherichia coli* K-12 Nramp orthologue (MntH) as a selective divalent metal ion transporter. Mol Microbiol 35:1065–1078

Martin DW, Baumgartner JE, Gee JM, Anderson ES, Roop RM II (2012) SodA is a major metabolic antioxidant in *Brucella abortus* 2308 that plays a significant, but limited, role in the virulence of this strain in the mouse model. Microbiology 158:1767–1774

Martin JE, Waters LS, Storz G, Imlay JA (2015) The *Escherichia coli* small protein MntS and exporter MntP optimize the intracellular concentration of manganese. PLoS Genet 11: e1004977

Martínez MR, Ugalde R, Almirón M (2005) Dimeric *Brucella abortus* Irr protein controls its own expression and binds haem. Microbiology 151:3427–3433

McAllister LJ, Tseng HJ, Ogunniyi AD, Jennings MP, McEwan AG, Paton JC (2004) Molecular analysis of the *psa* permease complex of *Streptococcus pneumoniae*. Mol Microbiol 53: 889–901

Mechold U, Cashel M, Steiner K, Gentry D, Malke H (1996) Functional analysis of a *relA/spoT* gene homolog from *Streptococcus equisimilis*. J Bacteriol 178:1401–1411

Menscher EA, Caswell CC, Anderson ES, Roop RM II (2012) Mur regulates the gene encoding the manganese transporter MntH in *Brucella abortus* 2308. J Bacteriol 194:561–566

Minder AC, de Rudder KEE, Narberhaus F, Fischer HM, Hennecke H, Geiger O (2001) Phosphatidylcholine levels in *Bradyrhizobium japonicum* membranes are critical for an efficient symbiosis with the soybean host plant. Mol Microbiol 39:1186–1198

Ouyang Z, He M, Oman T, Yang XF, Norgard MV (2009) A manganese transporter, BB0219 (BmtA), is required for virulence of the Lyme disease spirochete, *Borrelia burgdorferi*. Proc Natl Acad Sci USA 106:3449–3454

Papp-Wallace KM, Maguire ME (2006) Manganese transport and the role of manganese in virulence. Annu Rev Microbiol 60:187–209

Patzer SI, Hantke K (2001) Dual repression by Fe^{2+}-Fur and Mn^{2+}-MntR of the *mntH* gene, encoding an NRAMP-like Mn^{2+} transporter in *Escherichia coli*. J Bacteriol 183:4806–4813

Perry RD, Craig SK, Abney J, Bobrov AG, Kirillina O, Mier I Jr, Trusczynska H, Fetherston JD (2012) Manganese transporters Yfe and MntH are Fur-regulated and important for the virulence of *Yersinia pestis*. Microbiology 158:804–815

Petersen E, Chaudhuri P, Gourley C, Harms J, Splitter G (2011) *Brucella melitensis* cyclic di-GMP phosphodiesterase BpdA controls expression of flagellar genes. J Bacteriol 193:5683–5691

Phillips RW, Seastone DJ, Norton DD, Farris MA, Kovach ME, Elzer PH, Baldwin CL, Roop RM II (1996) Molecular cloning of the *Brucella abortus* iron uptake regulation (*fur*) gene. In: Abstracts of the 96th Annual Meeting of the American Society for Microbiology, Abstract D-24, p 245

Platero RA, Jaureguy M, Battistoni FK, Fabiano ER (2003) Mutations in *sitB* and *sitD* genes affect manganese-growth requirements in *Sinorhizobium meliloti*. FEMS Microbiol Lett 218:65–70

Platero R, Peixoto L, O'Brian MR, Fabiano E (2004) Fur is involved in manganese-dependent regulation of *mntA* (*sitA*) expression in *Sinorhizobium meliloti*. Appl Environ Microbiol 70:4349–4355

Que Q, Helmann JD (2000) Manganese homeostasis in *Bacillus subtilis* is regulated by MntR, a bifunctional regulator related to the diphtheria toxin repressor family of proteins. Mol Microbiol 35:1454–1468

Raimunda D, Elso-Berberián G (2014) Functional characterization of the CDF transporter SMc02724 (SmYiiP) in *Sinorhizobium meliloti*: roles in manganese homeostasis and nodulation. Biochim Biophys Acta 1838:3203–3211

Robertson DC, McCullough WG (1968a) The glucose catabolism of the genus *Brucella*. I. Evaluation of pathways. Arch Biochem Biophys 127:263–273

Robertson DC, McCullough WG (1968b) The glucose catabolism of the genus *Brucella*. II. Cell-free studies with *B. abortus* (S-19). Arch Biochem Biophys 127:445–456

Rodionov DA, Gelfand MS, Todd JD, Curson ARJ, Johnston AWB (2006) Computational reconstruction of iron- and manganese-responsive transcriptional networks in α-proteobacteria. PLoS Comput Biol 2:1568–1585

Römling U, Galperin MY, Gomelsky M (2013) Cyclic di-GMP: the first 25 years of a universal bacterial second messenger. Microbiol Mol Biol Rev 77:1–52

Roop RM II, Gaines JM, Anderson ES, Caswell CC, Martin DW (2009) Survival of the fittest: how *Brucella* strains adapt to their intracellular niche in the host. Med Microbiol Immunol 198:221–238

Rosch JW, Gao G, Ridout G, Yang YD, Toumanen EI (2009) Role of the manganese efflux system *mntE* for signalling and pathogenesis in *Streptococcus pneumoniae*. Mol Microbiol 72:12–25

Ryan RP (2013) Cyclic di-GMP signalling and the regulation of bacterial virulence. Microbiology 159:1286–1297

Sanders TH, Higuchi K, Brewer CR (1953) Studies on the nutrition of *Brucella melitensis*. J Bacteriol 66:294–299

Silver S, Johnseine P, Whitney E, Clark D (1972) Manganese-resistant mutants of *Escherichia coli*: physiologic and genetic studies. J Bacteriol 110:186–195

Slade D, Radman M (2011) Oxidative stress resistance in *Deinococcus radiodurans*. Microbiol Mol Biol Rev 75:133–191

Sobota JM, Imlay JA (2011) Iron enzyme ribulose-5-phosphate 3-epimerase in *Escherichia coli* is rapidly damaged by hydrogen peroxide but can be protected by manganese. Proc Natl Acad Sci U S A 108:5402–5407

Sohlenkamp C, López-Lara IM, Geiger O (2003) Biosynthesis of phosphatidylcholine in bacteria. Prog Lipid Res 42:115–162

Sriranganathan N, Boyle SM, Schurig G, Misra H (1991) Superoxide dismutases of virulant and avirulent strains of *Brucella abortus*. Vet Microbiol 26:359–366

Stabel TJ, Sha Z, Mayfield JE (1994) Periplasmic location of *Brucella abortus* Cu/Zn superoxide dismutase. Vet Microbiol 38:307–314

Steele KH, Baumgartner JE, Valderas MW, Roop RM II (2010) Comparative study of the roles of AhpC and KatE as respiratory antioxidants in *Brucella abortus* 2308. J Bacteriol 192: 4912–4922

Tamayo R, Tischler AD, Camilli A (2005) The EAL domain protein VieA is a cyclic diguanylate phosphodiesterase. J Biol Chem 280:33324–33330

Turner AG, Ong CL, Gillen CM, Davies MR, West NP, McEwan AG, Walker MJ (2015) Manganese homeostasis in Group A *Streptococcus* is critical for resistance to oxidative stress and virulence. MBio 6:e00278–15

Veyrier FJ, Boneca IG, Cellier MF, Taha MK (2011) A novel metal transporter mediating manganese export (MntX) regulates the Mn to Fe intracellular ratio and *Neisseria meningtidis* virulence. PLoS Pathog 7:e1002261

Vidal SM, Malo D, Vogan K, Skamene E, Gros P (1993) Natural resistance to infection with intracellular parasites: isolation of a candidate for *Bcg*. Cell 73:469–485

Waldron KJ, Robinson NJ (2009) How do bacterial cells ensure that metalloproteins get the correct metal? Nat Rev Microbiol 6:25–35

Wang X, Tong H, Dong X (2014) PerR-regulated manganese ion uptake contributes to oxidative stress defense in an oral streptococcus. Appl Environ Microbiol 80:2351–2359

Waters LS, Sandoval M, Storz G (2011) The *Escherichia coli* MntR miniregulon includes genes encoding a small protein and an efflux pump required for manganese homeostasis. J Bacteriol 193:5887–5897

Wells DH, Long SR (2002) The *Sinorhizobium meliloti* stringent response affects multiple aspects of symbiosis. Mol Microbiol 43:1115–1127

Wessel M, Klüsener S, Gödeke J, Fritz C, Hacker S, Narberhaus F (2006) Virulence of *Agrobacterium tumefaciens* requires phosphatidylcholine in the bacterial membrane. Mol Microbiol 62:906–915

Xavier MN, Winter MG, Spees AM, den Hartigh AB, Nguyen K, Roux CM, Silva TMA, Atluri VL, Kerrinnes T, Keestra AM, Monack DM, Luciw PA, Eigenheer RA, Bäumler AJ, Santos RL, Tsolis RM (2013) PPARγ-mediated increase in glucose availability sustains chronic *Brucella abortus* infection in alternative activated macrophages. Cell Host Microbe 14:159–170

Young KJ, Brennan BJ, Tagore R, Brudvig GW (2015) Photosynthetic water oxidation: insights from manganese model chemistry. Acc Chem Res 48:567–574

Zaharik ML, Cullen VL, Fung AM, Libby SJ, Choy SLK, Coburn B, Kehres DG, Maguire ME, Fang FC, Finlay BB (2004) The *Salmonella enterica* serovar Typhimurium divalent cation transport systems MntH and SitABCD are essential for virulence in an *Nramp1*[G169] murine typhoid model. Infect Immun 72:5522–5525

Chapter 4
The Role of Zinc in the Biology and Virulence of *Brucella* Strains

Clayton C. Caswell

Abstract Zinc is an essential micronutrient for many bacteria, including *Brucella* spp., and in *Brucella* strains, zinc is required for the proper function of several important enzymes involved in amino acid biosynthesis and resistance to oxidative stress. Additionally, zinc is an important cofactor for the virulence-associated transcriptional regulatory protein MucR, as well as for RicA, a type IV secretion system effector protein. Owing to the importance of zinc in the biology of *Brucella* strains, these bacteria possess a high affinity zinc uptake system, called ZnuABC that preferential imports zinc. Conversely, because of the toxic potential of free zinc cations, the brucellae also encode a zinc export system, called ZntA that functions in the intracellular detoxification of excess zinc. Not surprisingly, the levels of the zinc uptake and export systems are precisely controlled, and this genetic regulation is performed by two different zinc-responsive regulatory proteins, called Zur and ZntR, that govern the expression of the *znuABC* and *zntA* genetic loci, respectively, at the level of transcription. Importantly, the proper homeostasis of zinc levels in *Brucella* is required for efficient pathogenesis in animal models of *Brucella* infection. This chapter will outline the systems that regulate zinc uptake and export in *Brucella* strains, including the role of these system in *Brucella* virulence; the genetic regulation of zinc homeostasis; and the function of zinc as a cofactor for important enzymes in *Brucella* strains.

Keywords Zinc · *Brucella* · SodC · MucR

C.C. Caswell (✉)
Department of Biomedical Sciences and Pathobiology, Virginia–Maryland College of Veterinary Medicine, Virginia Tech, Blacksburg, VA 24061, USA
e-mail: caswellc@vt.edu

© Springer International Publishing AG 2017
R. Martin Roop II and Clayton C. Caswell (eds.), *Metals and the Biology and Virulence of Brucella*, DOI 10.1007/978-3-319-53622-4_4

4.1   Zinc Import by ZnuABC

Brucella spp. encode a single high affinity zinc import system at the *znuABC* locus (Fig. 4.1). In many bacteria, *znuABC* genes encode a prototypical periplasmic-binding protein, ZnuA, that functions in concert with the inner membrane protein ZnuB and the ATPase protein ZnuC to facilitate the uptake of zinc into the bacterial cell (Hantke 2005). In *Brucella* spp., it has been demonstrated that ZnuA plays a crucial role in zinc homeostasis. Indeed, deletion of *znuA* in *B. abortus* led to increased sensitivity of the *znuA* mutant to the chelator EDTA, but this defect could be rescued by the addition of excess zinc chloride (Kim et al. 2004). Moreover, work by Yang and colleagues revealed that a *B. abortus znuA* mutant was unable to grow in the absence of zinc, supporting the proposition that ZnuA is an essential component of the zinc uptake system in *Brucella* (Yang et al. 2006). To date, there is no empirical data available describing the role of ZnuB and ZnuC in zinc import, but given the synteny of the *znuBC* genes with *znuA*, it is predicted that these proteins, along with ZnuA, are the cognate proteins needed for zinc uptake in *Brucella*.

In regards to pathogenesis, it is important to note that deletion of *znuA* results in attenuation of *Brucella* strains (Table 4.1). *Brucella znuA* mutants are defective in their capacity to survive and replicate in HeLa epithelial cells, primary murine bone marrow-derived macrophages, and RAW 264.7 murine macrophages (Kim et al. 2004; Yang et al. 2006; Clapp et al. 2011). In vivo colonization studies demonstrated that *Brucella znuA* mutants are also attenuated in experimentally infected BALB/c mice (Kim et al. 2004; Yang et al. 2006; Clapp et al. 2011). Moreover, a *B. abortus znuA* mutant showed effectiveness as a live, attenuated vaccine strain, as

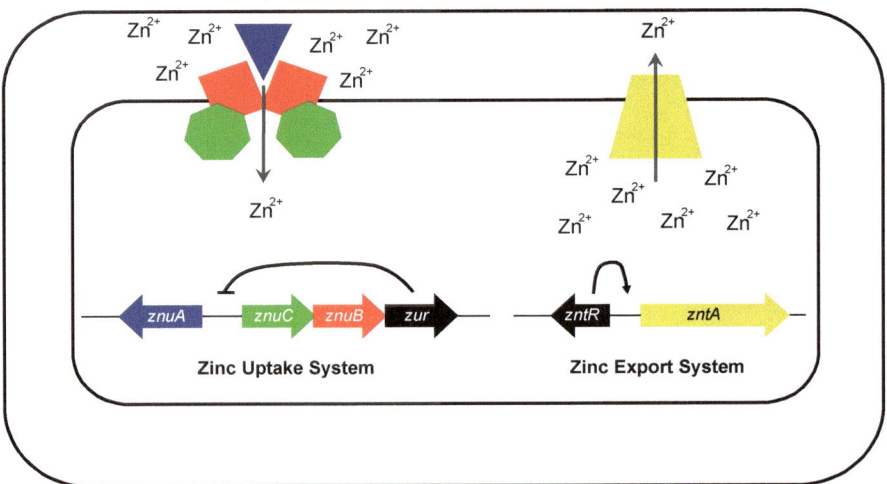

Fig. 4.1 The systems involved in zinc uptake (i.e. Znu) and export (i.e. Znt) in *Brucella*, and the genetic regulation of these systems

Table 4.1 Zinc-related genes and connections to virulence in *Brucella*

Gene	Locus tag[a]	Description of attenuation exhibited by mutant	Reference
znuA	BAB2_1079 (BAB_RS31485)	Attenuated in cells and mice	Kim et al. (2004), Yang et al. (2006), Clapp et al. (2011)
znuB	BAB2_1081 (BAB_RS31495)	No data currently available	N/A
znuC	BAB2_1080 (BAB_RS31490)	No data currently available	N/A
zur	BAB2_1082 (BAB_RS31500)	Not attenuated in primary murine macrophages or in mice	Sheehan et al. (2015)
zntA	BAB1_2019 (BAB_RS25545)	Not attenuated in primary murine macrophages or in mice	Sheehan et al. (2015)
zntR	BAB1_2018 (BAB_RS25540)	Not attenuated in primary murine macrophages; Attenuated in mice	Sheehan et al. (2015)
sodC	BAB2_0535 (BAB_RS28905)	No attenuated in HeLa cells or J774.A1 macrophages; Attenuated in primary murine macrophages and in mice	Tatum et al. (1992), Gee et al. (2005)
mucR	BAB1_0594 (BAB_RS18765)	Attenuated in primary, RAW264.7, and J774.A1 macrophages, and in mice	Caswell et al. (2013), Dong et al. (2013), Mirabella et al. (2013)
ricA	BAB1_1279 (BAB_RS22040)	Not attenuated in HeLa cells, bovine macrophages, or in mice	de Barsy et al. (2011)
bab1_0270	BAB1_0270 (BAB_RS17205)	Attenuated in HeLa cells, J774.A1 macrophages, and in mice	Ortiz-Román et al. (2014)
hisD	BAB1_0285 (BAB_RS17285)	Attenuated in THP-1 macrophages	Köhler et al. (2002)
bab1_1837	BAB1_1837 (BAB_RS24650)	No data currently available	N/A
bab2_0449	BAB2_0449 (BAB_RS28510)	No data currently available	N/A

[a]Locus tags correspond to the *Brucella melitensis* biovar Abortus 2308 genome sequence and annotation. The old (i.e., BAB1/2_####) and new (BAB_RS#####) locus tag designations are provided

intraperitoneal vaccination of BALB/c mice with the *znuA* mutant resulted in significant decreases in brucellae colonization of the spleens of mice following intraperitoneal challenge with wild-type *B. abortus* 2308 (Yang et al. 2006). In another study, it was shown that mice orally vaccinated with a *B. melitensis znuA* mutant strain were protected from subsequent nasal challenge with wild-type

B. melitensis 16M (Clapp et al. 2016). Taken together, the current data reveal that the ZnuABC system is important for zinc homeostasis and virulence in *Brucella* spp.

4.2 Zinc Export by ZntA

Given the potential toxicity of free zinc cations, it is not surprising that bacteria have evolved systems to purge this toxic atom in the event that excess zinc begins to accumulate intracellularly. Many bacteria employ ZntA proteins for the efflux and detoxification of zinc, and these transporters are P-type ATPases that function to remove zinc from the bacterial cell (Hantke 2001; Rosen 2002). Similarly, the high affinity expulsion of zinc from the *Brucella* cell is mediated by a ZntA protein that is encoded at the *zntA* locus (Fig. 4.1). ZntA is required for the resistance of *Brucella* strains to zinc stress, and this was demonstrated using a *B. abortus zntA* deletion strain (Sheehan et al. 2015). In the study by Sheehan and colleagues, the *zntA* deletion strain was significantly more sensitive to zinc-mediated killing compared to the wild-type strain *B. abortus* 2308. It was also determined that deletion of *zntA* did not lead to a generalized sensitivity to heavy metal stress, as there was no difference between the wild-type strain and the *zntA* deletion strain to withstand toxic levels of nickel or copper cations. Interestingly, deletion of *zntA* did not diminish the ability of the strain to survive and replicate in primary murine peritoneal macrophages, and the *zntA* mutant strain was not defective in its capacity to infect and colonize the spleens of intraperitoneally infected BALB/c mice. Therefore, while it has been clearly demonstrated that ZntA plays a key role in the in vitro detoxification of zinc in *Brucella* in the laboratory, ZntA is dispensable for the ability of *B. abortus* to infect mice. However, it is possible that ZntA is required for *Brucella* infection in the natural host(s), but more work is needed to assess the role of ZntR in these natural environments.

4.3 Zinc-Responsive Transcriptional Regulators Zur and ZntR

The control of zinc uptake and efflux must be precise and coordinated, because unregulated fluctuations in intracellular zinc concentrations could wreak havoc on the bacterium (Waldron and Robinson 2009). The primary means of controlling zinc uptake in *Brucella* is via the zinc uptake regulator, Zur, that is encoded at the *znuABC* locus, downstream of *znuBC* (Fig. 4.1). Zur is a member of the Fur-family of transcriptional regulatory proteins that control the expression of genes related to metal homeostasis, most notably iron homeostasis (Troxell and Hassan 2013). Experimental evidence has demonstrated that Zur functions in a zinc-responsive manner to control the expression of *znuABC* in *Brucella* (Sheehan et al. 2015). A *B.*

abortus zur deletion strain exhibited significantly increased levels of *znuA*, *znuB*, and *znuC* mRNA when the strain was grown in zinc-replete medium, indicating that Zur is a transcriptional repressor in *B. abortus*. Moreover, repression of *znuABC* expression by Zur is mediated by direct interactions between Zur and the DNA promoter region of the *znuABC* locus. This was demonstrated using electrophoretic mobility shift assays (EMSAs). Importantly, the interactions between Zur and the *znuABC* promoter are zinc-dependent, as chelation of divalent metal cations abolished binding between Zur and the *znuABC* promoter region DNA, and the supplementation of zinc, but not other divalent metal cations, restored the binding of Zur to the *znuABC* promoter. Regarding virulence, deletion of *zur* did not affect the ability of *Brucella* to survive and replicate in primary murine peritoneal macrophages nor to produce chronic spleen infection in BALB/c mice. As with the situation with *zntA* described above, it is possible that Zur is required for *Brucella* infections in natural host animals, but additional work is required to address this possibility.

Similar to the control of the zinc import system at the level of transcription, expression of the gene encoding the ZnuA efflux protein is regulated by a zinc-responsive transcriptional regulatory protein. This protein, called ZntR, is a member of the MerR family of regulators, and these proteins mediate transcriptional activation via a mechanism of DNA bending that "opens" the promoter of the regulated gene, thereby allowing RNA polymerase access to the promoter and subsequent initiation of transcription (Ansari et al. 1995; Hobman et al. 2005). Therefore, the basil activity of ZntR is repressive, but upon binding of a zinc cation, ZntR contorts the DNA to allow active transcription of the *zntA* gene. In *Brucella*, *zntR* is encoded divergent to *zntA* (Fig. 4.1), and the *Brucella* ZntR protein functions as a zinc-responsive transcriptional activator of *zntA* expression (Sheehan et al. 2015). Specifically, a *B. abortus zntA* deletion strain exhibited elevated levels of *zntA* mRNA, indicating that the *Brucella* ZntR activates transcription according to the paradigm for MerR-type transcriptional activators (Ansari et al. 1995). The work by Sheehan and colleagues also demonstrated that the *Brucella* ZntR protein binds directly to the *zntA* promoter, but this binding is not dependent upon zinc, as the addition of chelators to EMSAs did not affect interactions between ZntR and the *zntA* promoter. It is likely that ZntR is bound to the promoter of *zntA* at all times, and when intracellular zinc levels are elevated, ZntR senses and responds to the zinc by bending the DNA and allowing for the expression of *zntA*. Importantly, deletion of *zntR* resulted in attenuation of the strain in a mouse model of *Brucella* infection, but the *zntR* mutant was not affected in its ability to survive and replicate in primary peritoneal macrophages. It was hypothesized that over-expression of *zntA* in the *zntR* deletion strain was responsible for this phenotype, because over-production of the ZnuA zinc exporter could, in effect, starve the brucellae for zinc during a chronic infection. To test this hypothesis, a strain lacking both *zntR* and *zntA* was constructed, and the *zntR zntA* double deletion strain exhibited wild-type levels of spleen colonization in the mouse model, thus supporting the hypothesis. Overall, the fine-tuning of zinc homeostasis in *Brucella* is controlled primarily at the level of transcription by uptake-specific (i.e. Zur) and efflux-specific (i.e. ZntR) regulatory proteins, and while these regulatory circuits have been identified and characterized in *Brucella*, more work is needed to fully understand these systems in the context of *Brucella* pathogenesis, particularly in natural hosts.

4.4   Zinc-Dependent Proteins in *Brucella*

Zinc homeostasis is essential for the ability of *Brucella* strains to infect the host, and moreover, *Brucella* strains must retain the ability to import zinc via the high-affinity ZnuABC system to successfully colonize animals (Kim et al. 2004; Yang et al. 2006; Clapp et al. 2011). This raises an interesting question: what proteins and other cell components require zinc in *Brucella*? In fact, there are several known and purported needs for zinc in the brucellae, including several proteins in *Brucella* that require a zinc cofactor for their optimal activity, and some of these zinc-dependent proteins are required for the full virulence of *Brucella*.

4.4.1   *The Cu-Zn Superoxide Dismutase C (SodC)*

SodC is a periplasmic superoxide dismutase (SOD) that is responsible for detoxi-fying exogenously produced superoxide (i.e., O_2^-) in the periplasmic space of many Gram-negative bacteria, including several pathogenic organisms (Battistoni 2003). Bacterial SodC proteins require copper and zinc cofactors for full enzymatic activity, where the zinc cation appears to play an important structural role, and because of this requirement for copper and zinc, the SodC protein is often referred to as Cu-Zn SOD (Bordo et al. 1994; Lynch and Kuramitsu 2000). In *Brucella*, SodC is required for the optimal resistance of the bacteria to exogenous O_2^-, as well as for the full virulence of the brucellae. It was demonstrated that a *B. abortus sodC* deletion strain was attenuated in BALB/c mice following intraperitoneal infection, but the *sodC* mutant strain exhibited a wild-type capacity to survive and replicate in HeLa epithelial cells and in J774.A1 macrophages (Tatum et al. 1992). In another study, Gee and colleagues determined that a *B. abortus sodC* deletion strain was more sensitive to O_2^- in vitro, and moreover, the *sodC* deletion strain was atten-uated in both primary murine peritoneal macrophages and in C57BL/6J mice infected intraperitoneally (Gee et al. 2005). These latter studies also employed an NADPH oxidase inhibitor to implicate SodC in protecting the intracellular brucellae from oxidative burst of macrophages. Altogether, the SodC protein represents an important zinc-utilizing component that is required for *Brucella* pathogenesis.

4.4.2 The Transcriptional Regulatory Protein MucR

Another critical protein that relies on zinc is the *Brucella* MucR transcriptional regulator. MucR-family proteins (also referred to as Ros-family proteins) are bacterial zinc-finger-containing transcriptional regulatory proteins found exclusively in the α-proteobacteria, and these transcriptional regulators require zinc for the structural integrity of the protein (Bouhouche et al. 2000; Malgieri et al. 2015). The *Brucella mucR* gene was first linked to virulence using a *mariner* transposon mutagenesis approach, where the mutagenesis screen revealed that the interruption of *mucR* in *B. melitensis* 16M leads to dramatic decreases in macrophage survival and in spleen colonization of mice (Wu et al. 2006). Subsequently, targeted *B. melitensis* and *B. abortus mucR* deletion strains were reported to be attenuated in a variety of cellular models, as well as in experimentally infected mice (Caswell et al. 2013; Dong et al. 2013; Mirabella et al. 2013). These studies also demonstrated that *Brucella mucR* mutants have a variety of dramatic phenotypes, including in vitro growth defects, membrane and LPS aberrations, and increased sensitivity to iron deprivation. Regarding gene regulation, *Brucella* MucR is linked to the expression of many important genes, including genes involved in flagellar expression, the type IV secretion system, and LPS modification, but the mechanism of MucR-mediated gene regulation remains to be fully elucidated (Caswell et al. 2013; Dong et al. 2013; Mirabella et al. 2013). Additionally, structural studies are needed to definitively characterize the role of zinc in the *Brucella* MucR protein.

4.4.3 The Type IV Secretion System Effector Protein RicA

The type IV secretion system (T4SS) encoded by the *virB* operon is a vital virulence determinant in *Brucella* spp., and several T4SS effector proteins have been identified in *Brucella* strains in recent years (Boschiroli et al. 2002; Ke et al. 2015). One of these T4SS effector proteins is RicA (Rab2 interacting conserved protein A), named for its ability to bind to the small GTPase Rab2 in eukaryotic cells (de Barsy et al. 2011). Structural analyses of RicA determined that this protein assumes a structure reminiscent of a γ-carbonic anhydrase, a class of enzymes that are known to require zinc for their enzymatic activity (Nkengfac et al. 2012; Herrou and Crosson 2013). Indeed, the crystalized RicA protein from *B. abortus* contained the zinc cation characteristic of γ-carbonic anhydrases, but RicA did not possess carbonic anhydrase activity in vitro (Herrou and Crosson 2013). While a *B. melitensis ricA* mutant strain exhibited altered intracellular trafficking compared to the wild-type strain in HeLa epithelial cells, there was no defect in the capacity of the *ricA* mutant to infect HeLa cells, bovine macrophages, or experimentally infected mice (de Barsy et al. 2011). Nonetheless, RicA is one of the few bona fide T4SS effector proteins in *Brucella* described to date, and zinc appears to be an important factor in the yet-to-be determined molecular activity of this protein.

4.4.4 Other Potential Zinc-Containing Proteins in Brucella

In addition to the proteins outlined above, there are other *Brucella* proteins predicted to contain or require zinc for their activity; however, in each case, there is very limited information to date about the molecular mechanisms of these proteins, as well as the exact role or requirement for zinc in each of these proteins. One such example is the *B. abortus* protein BAB1_0270, which is predicted to function as a zinc-dependent metallopeptidase. Strikingly, a *B. abortus babl_0270* deletion strain was diminished in its ability to survive and replicate in HeLa epithelial cells and in J774.A1 macrophages compared to the wild-type strain, and the *babl_0270* was also attenuated in a mouse model of infection (Ortiz-Román et al. 2014). Future work will be required to fully characterize BAB1_0270 at the molecular level, particularly in regards to the role of zinc in its structure and/or function.

The protein HisD is a histidinol dehydrogenase that catalyzes the final two steps in the biosynthesis of histidine in bacteria, and HisD is a metalloprotein that require a zinc cation for proper enzymatic activity (Alifano et al. 1996). Köhler and colleagues revealed that a *B. suis hisD* mutant exhibited defective intracellular multiplication in THP-1 macrophages, and subsequent work has demonstrated that the targeting of HisD with small molecule inhibitors impedes the growth of *Brucella* in culture, as well as intracellularly growing brucellae in THP-1 macrophages (Köhler et al. 2002; Lopez et al. 2012). However, to date, there is no experimental data describing the requirement of zinc by HisD from *Brucella* spp.

Another class of zinc-containing enzymes is the carbonic anhydrases, which catalyze the hydration of carbon dioxide, thereby producing bicarbonate and protons, and the zinc cofactor contained in carbonic anhydrases is essential for the enzymatic activity of these proteins (Smith and Ferry 2000). *B. abortus* encodes two carbonic anhydrases at loci *babl_1837* and *bab2_0449*, and it has been reported that small molecule targeting of the BAB1_1837 and BAB2_0449 orthologs in *B. suis* 1330 (i.e., BR1829 and BRA0788, respectively) can inhibit growth of the brucellae in culture medium (Joseph et al. 2010, 2011; Lopez et al. 2012; Maresca et al. 2012). However, no studies have been performed with targeted *babl_1837* and *bab2_0449* deletion strains to assess the role of the enzymes encoded by these genes on the virulence of *Brucella*, and similarly, no information is currently available regarding the role of zinc in BAB1_1837 and BAB2_0449.

Acknowledgements The Caswell laboratory is supported by grants from the American Heart Association (15SDG23280044) and the National Institute of Allergy and Infectious Diseases (AI117648), and by internal support provided by the VA-MD College of Veterinary Medicine at Virginia Tech.

References

Alifano P, Fani R, Liò P, Lazcano A, Bazzicalupo M, Carlomagno MS, Bruni CB (1996) Histidine biosynthesis pathway and genes: structure, regulation, and evolution. Microbiol Rev 60:44–69

Ansari AZ, Bradner JE, O'Halloran TV (1995) DNA-bend modulation in a repressor-to-activator switching mechanism. Nature 374:371–375

Battistoni A (2003) Role of prokaryotic Cu, Zn superoxide dismutase in pathogenesis. Biochem Soc Trans 31:1326–1329

Bordo D, Djinovic K, Bolognesi M (1994) Conserved patterns in the Cu, Zn superoxide dismutase family. J Mol Biol 238:366–386

Boschiroli ML, Ouahrani-Bettache S, Foulongne V, Michaux-Charachon S, Bourg G, Allardet-Servent A, Cazevieille C, Lavigne JP, Liautard JP, Ramuz M, O'Callaghan D (2002) Type IV secretion and *Brucella* virulence. Vet Microbiol 90:341–348

Bouhouche N, Syvanen M, Kado CI (2000) The origin of prokaryotic C2H2 zinc finger regulators. Trends Microbiol 8:77–81

Caswell CC, Elhassanny AE, Planchin EE, Roux CM, Weeks-Gorospe JN, Ficht TA, Dunman PM, Roop RM II (2013) Diverse genetic regulon of the virulence-associated transcriptional regulator MucR in *Brucella abortus* 2308. Infect Immun 81:1040–1051

Clapp B, Skyberg JA, Yang X, Thornburg T, Walters N, Pascual DW (2011) Protective live oral brucellosis vaccines stimulate Th1 and th17 cell responses. Infect Immun 79:4165–4174

de Barsy M, Jamet A, Filopon D, Nicolas C, Laloux G, Rual JF, Muller A, Twizere JC, Nkengfac B, Vandenhaute J, Hill DE, Salcedo SP, Gorvel JP, Letesson JJ, De Bolle X (2011) Identification of a *Brucella* spp. secreted effector specifically interacting with human small GTPase Rab2. Cell Microbiol 13:1044–1058

Dong H, Liu W, Peng X, Jing Z, Wu Q (2013) The effects of MucR on expression of type IV secretion system, quorum sensing system and stress responses in *Brucella melitensis*. Vet Microbiol 166:535–542

Gee JM, Valderas MW, Kovach ME, Grippe VK, Robertson GT, Ng W-L, Richardson JM, Winkler ME, Roop RM II (2005) The Brucella abortus Cu, Zn superoxide dismutase is required for optimal resistance to oxidative killing by murine macrophages and wild-type virulence in experimentally infected mice. Infect Immun 73:2873–2880

Hantke K (2001) Bacterial zinc transporters and regulators. Biometals 14:239–249

Hantke K (2005) Bacterial zinc uptake and regulators. Curr Opin Microbiol 8:196–202

Herrou J, Crosson S (2013) Molecular structure of the *Brucella abortus* metalloprotein RicA, a Rab2-binding virulence effector. Biochemistry 52:9020–9028

Hobman JL, Wilkie J, Brown NL (2005) A design for life: prokaryotic metal-binding MerR family regulators. Biometals 18:429–436

Joseph P, Turtaut F, Ouahrani-Bettache S, Montero JL, Nishimori I, Minakuchi T, Vullo D, Scozzafava A, Köhler S, Winum JY, Supuran CT (2010) Cloning, characterization, and inhibition studies of a beta-carbonic anhydrase from *Brucella suis*. J Med Chem 53:2277–2285

Joseph P, Ouahrani-Bettache S, Montero JL, Nishimori I, Minakuchi T, Vullo D, Scozzafava A, Winum JY, Köhler S, Supuran CT (2011) A new β-carbonic anhydrase from *Brucella suis*, its cloning, characterization, and inhibition with sulfonamides and sulfamates, leading to impaired pathogen growth. Bioorg Med Chem 19:1172–1178

Ke Y, Wang Y, Li W, Chen Z (2015) Type IV secretion system of *Brucella* spp. and its effectors. Front Cell Infect Microbiol 5:72

Kim S, Watanabe K, Shirahata T, Watarai M (2004) Zinc uptake system (*znuA* locus) of *Brucella abortus* is essential for intracellular survival and virulence in mice. J Vet Med Sci 66:1059–1063

Köhler S, Foulongne V, Ouahrani-Bettache S, Bourg G, Teyssier J, Ramuz M, Liautard JP (2002) The analysis of the intramacrophagic virulome of *Brucella suis* deciphers the environment encountered by the pathogen inside the macrophage cell. Proc Natl Acad Sci USA 99:15711–15716

Lopez M, Köhler S, Winum JY (2012) Zinc metalloenzymes as new targets against the bacterial pathogen *Brucella*. J Inorg Biochem 111:138–145

Lynch M, Kuramitsu H (2000) Expression and role of superoxide dismutase (SOD) in pathogenic bacteria. Microbes Infect 2:1245–1255

Malgieri G, Palmieri M, Russo L, Fattorusso R, Pedone PV, Isernia C (2015) The prokaryotic zinc-finger: structure, function, and comparison with the eukaryotic counterpart. FEBS J 282:4480–4496

Maresca A, Scozzafava A, Köhler S, Winum JY, Supuran CT (2012) Inhibition of beta-carbonic anhydrases from the bacterial pathogen *Brucella suis* with inorganic anions. J Inorg Biochem 110:36–39

Mirabella A, Terwagne M, Zygmunt MS, Cloeckaert A, De Bolle X, Letesson JJ (2013) *Brucella melitensis* MucR, an orthologue of *Sinorhizobium meliloti* MucR, is involved in resistance to oxidative, detergent, and saline stresses and cell envelope modifications. J Bacteriol 195:453–465

Nkengfac B, Pouyez J, Bauwens E, Vandenhaute J, Letesson JJ, Wouters J, De Bolle X (2012) Structural analysis of *Brucella abortus* RicA substitutions that do not impair interaction with human Rab2 GTPase. BMC Biochem 13:16

Ortiz-Román L, Riquelme-Neira R, Roberto V, Oñate A (2014) Roles of genomic island 3 (GI-3) BAB1_0267 and BAB1_0270 open reading frames (ORFs) in the virulence of *Brucella abortus* 2308. Vet Microbiol 172:279–284

Rosen BP (2002) Transport and detoxification systems for transition metals, heavy metals and metalloids in eukaryotic and prokaryotic microbes. Comp Biochem Physiol A Mol Integr Physiol 133:689–693

Sheehan LM, Budnick JA, Roop RM II, Caswell CC (2015) Coordinated zinc homeostasis is essential for the wild-type virulence of *Brucella abortus*. J Bacteriol 197:1582–1591

Smith KS, Ferry JG (2000) Prokaryotic carbonic anhydrases. FEMS Microbiol Rev 24:335–366

Tatum FM, Detilleux PG, Sacks JM, Halling SM (1992) Construction of Cu-Zn superoxide dismutase deletion mutants of *Brucella abortus*: analysis of survival in vitro in epithelial and phagocytic cells and in vivo in mice. Infect Immun 60:2863–2869

Troxell B, Hassan HM (2013) Transcriptional regulation by Ferric Uptake Regulator (Fur) in pathogenic bacteria. Front Cell Infect Microbiol 3:59

Waldron KJ, Robinson NJ (2009) How do bacterial cells ensure that metalloproteins get the correct metal? Nat Rev Microbiol 7:25–35

Wu Q, Pei J, Turse C, Ficht TA (2006) Mariner mutagenesis of *Brucella melitensis* reveals genes with previously uncharacterized roles in virulence and survival. BMC Microbiol 6:102

Yang X, Becker T, Walters N, Pascual DW (2006) Deletion of *znuA* virulence factor attenuates *Brucella abortus* and confers protection against wild-type challenge. Infect Immun 74:3874–3879

Chapter 5
Nickel Homeostasis in *Brucella* spp.

James A. Budnick and Clayton C. Caswell

Abstract Nickel is an important cofactor for microbial proteins, such as urease, that are involved in the adaptation of bacteria to stressful conditions, as well as other proteins related to general metabolism. Therefore, successful acquisition of nickel from the environment is essential for microbes to survive. Like most metals though, acquisition of nickel is a double edge sword, as high intracellular concentrations of nickel can be toxic to microbes. Thus, bacteria have developed ways to tightly control intracellular nickel concentrations. Much of what is known about the mechanisms of nickel uptake, export, and regulation have been determined in *Escherichia coli* and *Helicobacter pylori*, but parallels between these systems and *Brucella* spp. can be drawn. This chapter will outline what is currently known about nickel acquisition by the NikABCDE and NikKMLQO systems, as well as propose the role of a putative nickel exporter and transcriptional regulators of genes encoding Ni import and export systems in *Brucella* biology and virulence.

Keywords *Brucella* · Nickel · Urease

5.1 Nickel Import by NikABCDE and NikKMLQO

With regards to nickel homeostasis in *Brucella*, nickel import has been the most studied system to date. *Brucella* spp. contain two loci encoding nickel import systems. NikABCDE is an ABC-type transporter that was first described in 2001 by Jubier-Maurin et al., and NikKLMQO is an energy coupling factor (ECF)-type transporter transcribed with the *ure2* operon (Sangari et al. 2010; Jubier-Maurin et al. 2001). *nikABCDE* encodes an archetypal ABC-type transporter with *nikD* and *nikE* encoding ATPases, *nikB* and *nikC* inner membrane proteins, and *nikA* as a periplasmic nickel-binding protein (Fig. 5.1). The functions of these proteins have

J.A. Budnick · C.C. Caswell (✉)
Department of Biomedical Sciences and Pathobiology, Virginia–Maryland College of
Veterinary Medicine, Virginia Tech, Blacksburg, VA 24061, USA
e-mail: caswellc@vt.edu

© Springer International Publishing AG 2017
R. Martin Roop II and Clayton C. Caswell (eds.), *Metals and the Biology and Virulence of Brucella*, DOI 10.1007/978-3-319-53622-4_5

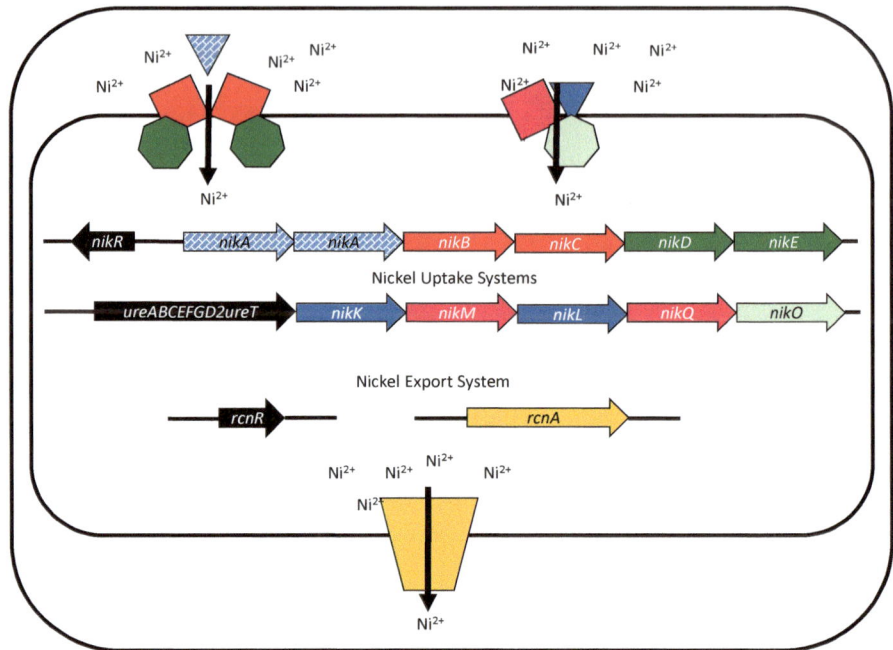

Fig. 5.1 Nickel uptake and export system

not been experimentally characterized in *Brucella* but can be inferred based on homology to other characterized nickel transporters (Rodionov et al. 2006). However, *B. suis* NikA has been crystallized and specific nickel binding sites were determined, adding to the evidence that NikA is the periplasmic binding protein of the *nikABCDE* system (Lebrette et al. 2014).

Utilizing a GFP-transcriptional fusion, it was shown that the *nikA* promoter in *B. suis* 1330 was induced during infection of J774.A1 macrophage-like cells, and similarly, the *nikA* promoter was activated when bacteria were grown in culture medium supplemented with EDTA or under microaerobic conditions (Jubier-Maurin et al. 2001). Despite being induced upon infection of J774.A1 cells, strains containing a partial deletion of *nikA* showed similar replication rates in both human monocytes and THP-1 cells, indicating that this system is not necessary for the survival and replication of *B. suis* during macrophage infection (Table 5.1).

Interestingly, the *nikABCDE* gene locus is dissimilar amongst *Brucella* spp. *nikD* is a pseudogene in *B. ovis* ATCC 25840 (Tsolis et al. 2009); and in *B. abortus* strains 2308, 9-941, and S19, there is a nonsense mutation about halfway through *nikA*, rendering it a pseudogene (Sangari et al. 2010). However, the presence of a second nickel import system likely compensates for these mutations in *B. ovis* and *B. abortus*.

While studying the *ure2* gene locus, Sangari et al. described the presence of an ECF-type nickel transporter gene cluster downstream of the *ure2* operon that encodes

Table 5.1 Nickel-related genes and connections to virulence in *Brucella*

Gene	Locus tag*	Description of phenotypes observed by mutant	Reference
nikA	BAB2_0433 and BAB2_0434	Not attenuated in THP-1 macrophages**	Jubier-Maurin et al. (2001)
nikB	BAB2_0435 (BAB_RS28430)	No data currently available	N/A
nikC	BAB2_0436 (BAB_RS28435)	No data currently available	N/A
nikD	BAB2_0437 (BAB_RS28440)	No data currently available	N/A
nikE	BAB2_0438 (BAB_RS28445)	No data currently available	N/A
nikK	BAB1_1384 (BAB_RS22535)	No data currently available	N/A
nikM	BAB1_1385 (BAB_RS22540)	No data currently available	N/A
nikL	BAB1_1386 (BAB_RS22545)	No data currently available	N/A
nikQ	BAB1_1387 (BAB_RS22550)	No data currently available	N/A
nikO	BAB1_1388 (BAB_RS22555)	Decreased urease activity and higher sensitivity to acidic pH stress	Sangari et al. (2010)
nikR	BAB2_0432 (BAB_RS28420)	Induced 12 hours post-infection of HeLa cells***	Rossetti et al. (2011)
rcnA	BAB2_0094 (BAB_RS26805)	No data currently available	N/A
rcnR	BAB1_0129 (BAB_RS16535)	No data currently available	N/A
*ureC*1	BAB1_0300 (BAB_RS17360)	Attenuated upon oral infection of BALB/c mice	Sangari et al. (2007)
*ureC*2	BAB1_1378 (BAB_RS22505)	Not attenuated upon oral infection of BALB/c mice	Sangari et al. (2007)

*Locus tags coorespond to *Brucella melitensis* biovar Abortus 2308 genome sequence and annotation. The old (i.e., BAB1/2_####) and new (BAB_RS#####) locus tag designation are provided

**Experiment was conducted using a deletion of *nikA* in *B. suis* 1330

***Experiment was conducted using *B. melitensis* 16M

the putative nickel transporter *nikKMLQO* (Sangari et al. 2010). ECF transporters lack a extracytoplasmic periplasmic binding protein and instead contain a membrane-embedded substrate binding protein (Wang et al. 2013; Erkens et al. 2011). Bioinformatic homology analyses predict *nikM* and *nikQ* to encode inner membrane proteins, *nikO* to encode an ATPase, and *nikK* and *nikL* to encode substrate-binding proteins (Fig. 5.1) (Sangari et al. 2010). Again, none of these functions have been experimentally demonstrated. Sangari et al. showed that the *ure*2 operon encodes 3 separate systems, *ureABCEFGD2*, *ureT*, and *nikKMLQO*. In their study, they constructed a *nikO* mutant to determine the role of this transporter on urease activity. It was shown that the *nikO* mutant had lower urease activity and was more sensitive to acidic pH in culture (Table 5.1). The *nikO* mutant was not used to infect cells in vitro.

Altogether, these data show that *Brucella* spp. contain two separately encoded nickel import systems, which likely compensate for one another. None of the above mutants were used to define the role of the *nik* genes in vivo. Further studies are necessary to characterize both import systems and to understand the potential for them to serve redundant functions in vivo. One method to do this would be to construct mutants containing mutations in each system individually and then construct a mutant containing mutations in both systems. These mutants could be used to elucidate the proposed redundancy of the systems and understand the role of nickel in *Brucella* virulence. As discussed below, the studies provide evidence that they are necessary for the import of nickel and for the efficient activity of urease during infection.

5.2   The Nickel and Cobalt Exporter RcnA

The expression of a nickel and cobalt export system, RcnA, is a mechanism utilized by bacteria to counter metal toxicity and was first described by Rodrigue et al. in 2005. That study demonstrated that *E. coli rcnA* was induced upon addition of nickel or cobalt to the media; strains containing a deletion of *rcnA* were more sensitive to nickel and cobalt toxicity and contained more intracellular nickel and cobalt; and strains harboring a multicopy plasmid expressing RcnA contained less intracellular nickel and cobalt than strains containing an empty multicopy plasmid (Rodrigue et al. 2005). *Brucella* spp. Encode a putative nickel and cobalt export permease RcnA (Fig. 5.1). However, *Brucella* RcnA shows less than 40% amino acid similarity to that of *E. coli* RcnA and is missing the distinctive histidine cluster found within *E. coli* RcnA sequence. This begs the question whether the ortholog of RcnA in *Brucella* is truly a nickel and cobalt exporter, or are there other mechanisms to detoxify the cells of these metal cations?

5.3   Nickel-Responsive Regulators NikR and RcnR

While nickel is an essential cofactor for several proteins in bacteria, an excessive amount of intracellular nickel can be toxic by causing oxidative damage, by replacing essential metal ions in metalloenzymes, or by binding to non-metalloenzyme active sites or secondary sites leading to decreased enzyme activity (Macomber and Hausinger 2011). To combat this problem, bacteria possess mechanisms to regulate intracellular nickel concentrations. *Escherichia coli* and *Helicobacter pylori* encode the well characterized nickel responsive regulator, NikR, which is a nickel dependent ribbon-helix-helix transcriptional regulator.

NikR regulates *nik* genes in response to intracellular metal concentrations and other stimuli, and *E. coli* also encodes a repressor of the nickel exporter RcnA, called RcnR (Iwig et al. 2006; Schreiter and Drennan 2007).

In *E. coli*, NikR solely regulates the *nik* operon (Li and Zamble 2009). In *H. pylori*, NikR is a repressor of the nickel uptake gene *nixA* and an activator of the *ure* operon (Ernst et al. 2005). A putative nickel responsive regulator is encoded adjacent to the *nikABCDE* locus in *B. abortus* 2308. The ability of NikR to regulate the nickel import systems and potentially the *ure2* operon has not been experimentally characterized, and there is little data describing the role of this regulator in *Brucella*. While the aim of the study was not to characterize NikR, Jubier-Maurin et al. identified a sequence upstream of *nikA* that closely resembles the NikR binding site in *E. coli* (Jubier-Maurin et al. 2001). Rossetti et al. were the first to identify a potential role of NikR in virulence during *B. melitensis* infection of HeLa cells. Their data demonstrated that *nikR* expression increased 12 h post-infection of HeLa cells infected with *B. melitensis* (Table 5.1) (Rossetti et al. 2011). This study also revealed increased expression of the *ure2* operon, but no differential expression of any of the *nik* genes during HeLa cell infection. Thus, the authors suggested that NikR could be both a repressor of the *nik* genes and an activator of the *ure2* genes (Rossetti et al. 2011). However, since this was not a targeted experiment, rather a discovery tool for genes differentially regulated during infection of HeLa cells, the link between NikR and the regulation of the *ure2* operon remains unknown. It is possible that another regulator is responsible for the induction of *ure2*, and it is coincidental that both *ure2* and *nikR* are induced upon infection of HeLa cells. It is clear that further studies are necessary to deduce the function of NikR in *Brucella* spp.

RcnR is a repressor of *rcnA* and *rcnR* in a nickel and cobalt responsive manner in *E. coli* (Blaha et al. 2011; Iwig et al. 2006). The two genes are transcribed divergently from one another in *E. coli*, and Blaha et al. identified a specific palindromic RcnR binding box, TACT-N7-AGTA, in the intergenic region of the two genes. Upon binding either of the metals, RcnR dissociated from the RcnR binding box, allowing for the expression of *rcnA* and *rcnR* (Blaha et al. 2011). Deletion of *rcnR* showed constitutively expressed *rcnA* (Iwig et al. 2006). *Brucella* spp. also encode a putative RcnR protein that is 40% identical and over 60% similar to the RcnR protein of *E. coli*. Contrary to the situation in *E. coli*, *rcnR* in *Brucella* strains is not located divergently to *rcnA,* but rather is located on a different chromosome from *rcnA*. Interestingly, an identical sequence to that of the *E. coli* RcnR binding box is located upstream of *rcnR*, but this putative RcnR-binding sequence is not observed upstream of *rcnA* in *Brucella*. It should be noted that *Brucella abortus* RcnR is 54% identical to the formaldehyde stress response regulator FrmR in *E. coli*, and the genomic organization of RcnR and surrounding genes in *B. abortus* is similar to that of the *E. coli frmR* and *N. gonorrhoeae nmlR* (an *fmrR* ortholog) loci (Chen et al. 2016). To date, no studies have characterized RcnR in *Brucella* strains, and empirical evidence will be needed to support the hypothesis that RcnR is a transcriptional regulator of *rcnR* and/or *rcnA* in *Brucella* spp. and not, in fact, an ortholog of FrmR related to formaldehyde resistance.

5.4 Nickel-Dependent Proteins in *Brucella*

Bacterial proteins that require nickel include urease, NiFe-hydrogenase, carbon monoxide dehydrogenase, acetyl-coenzyme A decarbonylase, methyl-coenzyme M reductase, nickel dependent superoxide dismutases and glyoxylases (Mulrooney and Hausinger 2003; Li and Zamble 2009). Of the above proteins, urease is the only protein directly linked to virulence in *Brucella* spp.

5.4.1 Urease

Urease enzymes hydrolyze urea into carbon dioxide and ammonia and thus, play a key role in nitrogen metabolism, as well as acid resistance as microbes pass through the acidic environment of the stomach (Li and Zamble 2009). Urease was one of the first proteins demonstrated to require nickel for catalysis (Alagna et al. 1984). Since then, extensive biochemical analyses of this protein have exposed much about specific binding sites for nickel within the protein (Mulrooney and Hausinger 2003). The *Brucella ure* operons can be split into two groups: structural proteins (encoded by *ureA*, *ureB*, and *ureC*) and accessory proteins (*ureD*, *ureE*, *ureF*, and *ureG*) (Sangari et al. 2007). This genetic organization is very similar to that of the genes encoding the trimeric urease of *Klebsiella aerogenes* (Mulrooney and Hausinger 1990; Sangari et al. 2007). It was shown that *K. aerogenes* UreE directly binds to nickel and is thought to function as a nickel carrier for the urease enzyme (Mulrooney et al. 2005). The *Brucella* genome contains two urease operons, *ure1* and *ure2*. While both have been shown to contribute to urease activity and are activated in acidic conditions, only *ure1* has been shown to be necessary for the virulence of *B. suis* or *B. abortus* in mice infected via the oral route (Bandara et al. 2007; Sangari et al. 2007). Upon deletion of either *ureC1* (*ureC* from operon 1) or *ureC2* (*ureC* from *ure* operon 2), *ureC1* was deemed less fit for oral infection of a mouse model compared to *B. abortus* 2308 (Sangari et al. 2007). This has been predicted to be due to the dissimilar genetic identities of the *ure* operons. In *B. suis* 1330, the amino acid similarity between *ure1* and *ure2* is about 70%, and most of the urease activity in this strain is predicted to be due to *ure1* (Bandara et al. 2007). Mutations in the *ure* genes does not affect the ability of *Brucella* spp. to survive and replicate in cell lines in vitro (Sangari et al. 2007; Bandara et al. 2007). This evidence supports the claim that *Brucella* is most likely utilizing urease to combat a low pH environment during the biologically relevant oral route of infection. Altogether, it has been shown that while *ure2* contributes to urease activity, it is not necessary for infection of a host via the oral route. Therefore, it is possible that *Brucella* spp. have maintained the *ure2* operon due to necessity of other genes (i.e., *ureT*, *nikK*, *nikM*, *nikL*, *nikQ*, *nikO*) located downstream of the *ure2* genes in the operon for infecting the host.

5.4.2 Other Potential Nickel-Containing Proteins in **Brucella**

As stated above, several proteins other than urease have been identified as nickel-binding metalloproteins in microbes. However, *Brucella* urease is the only nickel binding protein that has been extensively studied in *Brucella* biology.

Acknowledgements Work in the Caswell laboratory is supported by grants from the American Heart Association (15SDG23280044) and the National Institute of Allergy and Infectious Diseases (AI117648), and by internal support provided by the VA-MD College of Veterinary Medicine at Virginia Tech.

References

Alagna L, Hasnain SS, Piggott B, Williams DJ (1984) The nickel ion environment in jack bean urease. Biochem J 220:591–595

Bandara AB, Contreras A, Contreras-Rodriguez A, Martins AM, Dobrean V, Poff-Reichow S, Rajasekaran P, Sriranganathan N, Schurig GG, Boyle SM (2007) *Brucella suis* urease encoded by *ure*1 but not *ure*2 is necessary for intestinal infection of BALB/c mice. BMC Microbiol 7:57

Blaha D, Arous S, Bleriot C, Dorel C, Mandrand-Berthelot MA, Rodrigue A (2011) The *Escherichia coli* metallo-regulator RcnR represses *rcnA* and *rcnR* transcription through binding on a shared operator site: insights into regulatory specificity towards nickel and cobalt. Biochimie 93:434–439

Chen NH, Djoko KY, Veyrier FJ, McEwan AG (2016) Formaldehyde stress responses in bacterial pathogens. Front Microbiol 7:257

Erkens GB, Berntsson RP, Fulyani F, Majsnerowska M, Vujicic-Zagar A, Ter Beek J, Poolman B, Slotboom DJ (2011) The structural basis of modularity in ECF-type ABC transporters. Nat Struct Mol Biol 18:755–760

Ernst FD, Kuipers EJ, Heijens A, Sarwari R, Stoof J, Penn CW, Kusters JG, van Vliet AH (2005) The nickel-responsive regulator NikR controls activation and repression of gene transcription in *Helicobacter pylori*. Infect Immun 73:7252–7258

Iwig JS, Rowe JL, Chivers PT (2006) Nickel homeostasis in *Escherichia coli*—the *rcnR-rcnA* efflux pathway and its linkage to NikR function. Mol Microbiol 62:252–262

Jubier-Maurin V, Rodrigue A, Ouahrani-Bettache S, Layssac M, Mandrand-Berthelot MA, Kohler S, Liautard JP (2001) Identification of the *nik* gene cluster of *Brucella suis*: regulation and contribution to urease activity. J Bacteriol 183:426–434

Lebrette H, Brochier-Armanet C, Zambelli B, de Reuse H, Borezee-Durant E, Ciurli S, Cavazza C (2014) Promiscuous nickel import in human pathogens: structure, thermodynamics, and evolution of extracytoplasmic nickel-binding proteins. Structure 22:1421–1432

Li Y, Zamble DB (2009) Nickel homeostasis and nickel regulation: an overview. Chem Rev 109:4617–4643

Macomber L, Hausinger RP (2011) Mechanisms of nickel toxicity in microorganisms. Metallomics 3:1153–1162

Mulrooney SB, Hausinger RP (1990) Sequence of the *Klebsiella aerogenes* urease genes and evidence for accessory proteins facilitating nickel incorporation. J Bacteriol 172:5837–5843

Mulrooney SB, Hausinger RP (2003) Nickel uptake and utilization by microorganisms. FEMS Microbiol Rev 27:239–261

Mulrooney SB, Ward SK, Hausinger RP (2005) Purification and properties of the *Klebsiella aerogenes* UreE metal-binding domain, a functional metallochaperone of urease. J Bacteriol 187:3581–3585

Rodionov DA, Hebbeln P, Gelfand MS, Eitinger T (2006) Comparative and functional genomic analysis of prokaryotic nickel and cobalt uptake transporters: evidence for a novel group of ATP-binding cassette transporters. J Bacteriol 188:317–327

Rodrigue A, Effantin G, Mandrand-Berthelot MA (2005) Identification of *rcnA* (*yohM*), a nickel and cobalt resistance gene in *Escherichia coli*. J Bacteriol 187:2912–2916

Rossetti CA, Galindo CL, Garner HR, Adams LG (2011) Transcriptional profile of the intracellular pathogen *Brucella melitensis* following HeLa cells infection. Microb Pathog 51:338–344

Sangari FJ, Cayon AM, Seoane A, Garcia-Lobo JM (2010) *Brucella abortus ure2* region contains an acid-activated urea transporter and a nickel transport system. BMC Microbiol 10:107

Sangari FJ, Seoane A, Rodriguez MC, Aguero J, Garcia Lobo JM (2007) Characterization of the urease operon of *Brucella abortus* and assessment of its role in virulence of the bacterium. Infect Immun 75:774–780

Schreiter ER, Drennan CL (2007) Ribbon-helix-helix transcription factors: variations on a theme. Nat Rev Microbiol 5:710–720

Tsolis RM, Seshadri R, Santos RL, Sangari FJ, Lobo JM, de Jong MF, Ren Q, Myers G, Brinkac LM, Nelson WC, Deboy RT, Angiuoli S, Khouri H, Dimitrov G, Robinson JR, Mulligan S, Walker RL, Elzer PE, Hassan KA, Paulsen IT (2009) Genome degradation in *Brucella ovis* corresponds with narrowing of its host range and tissue tropism. PLoS ONE 4: e5519

Wang T, Fu G, Pan X, Wu J, Gong X, Wang J, Shi Y (2013) Structure of a bacterial energy-coupling factor transporter. Nature 497:272–276

Chapter 6
Magnesium, Copper and Cobalt

R. Martin Roop II, John E. Baumgartner, Joshua E. Pitzer
and Daniel W. Martin

Abstract Magnesium, copper and cobalt are essential micronutrients for *Brucella* strains, but relatively little is known about how the brucellae acquire the levels of these metals they need and avoid their toxicity. This chapter will review the information that is available in the literature and can be derived from surveys of currently available genome sequences regarding magnesium, copper and cobalt homeostasis in *Brucella*.

Keywords *Brucella* · Magnesium · Copper · Cobalt · Cobalamin

6.1 Magnesium

Magnesium (Mg) is the most abundant divalent cation in bacterial cells (Papp-Wallace and Maguire 2008; Groisman et al. 2013). Because it sits at the bottom of the Irving-Williams scale of metal reactivity, cells can accumulate very high (e.g. mM) concentrations of Mg without experiencing toxicity (Waldron and Robinson 2009) and the intracellular concentrations of this metal are typically 1000-fold higher than that of other metals (Foster et al. 2014). Because of its positive charge, Mg^{2+} neutralizes the negative charges in phosphate groups and other anions, and plays an important role in the stabilization of nucleic acids, phospholipid membranes and ribosomes. It is also the most common metal co-factor in enzymes (Andreini et al. 2008).

R.M. Roop II (✉) · J.E. Baumgartner · J.E. Pitzer · D.W. Martin
Department of Microbiology and Immunology, Brody School of Medicine,
East Carolina University, Greenville, NC 27834, USA
e-mail: roopr@ecu.edu

© Springer International Publishing AG 2017
R. Martin Roop II and Clayton C. Caswell (eds.), *Metals and the Biology and Virulence of Brucella*, DOI 10.1007/978-3-319-53622-4_6

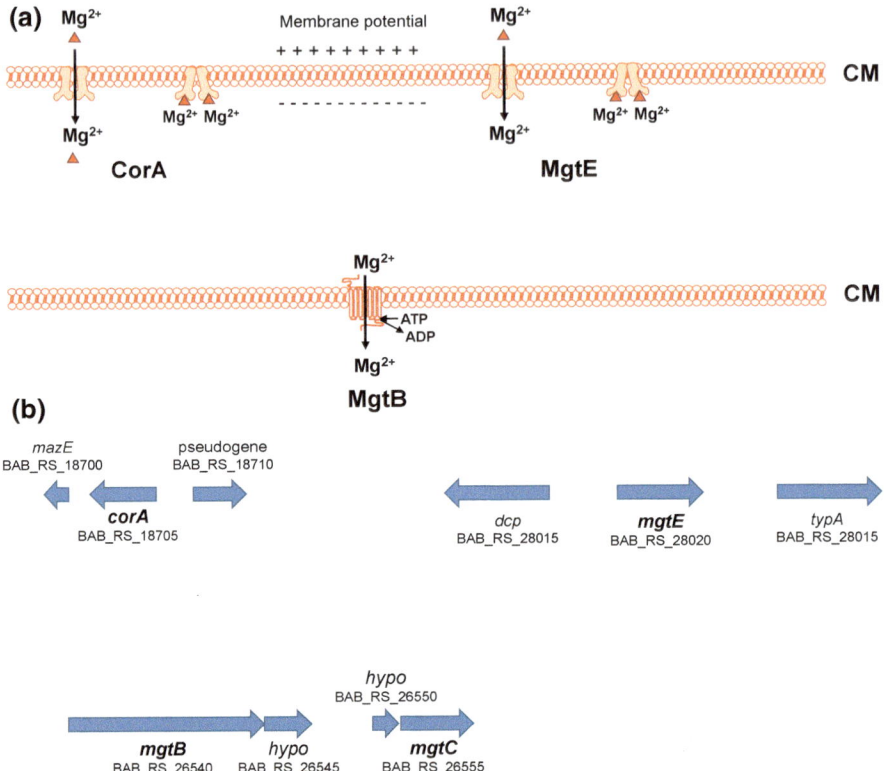

Fig. 6.1 **a** Proposed roles of the *Brucella* CorA, MgtE and MgtB in Mg^{2+} transport, and **b** genetic organization of the corresponding genes. The gene designations are those used in the *B. abortus* 2308 genome sequence. CM—cytoplasmic membrane

6.1.1 CorA, MgtE and MgtB

Mg represents an essential micronutrient for *Brucella* strains during in vitro culture (Sanders et al. 1953; Evenson and Gerhardt 1955), but a detailed examination of the cellular components that control Mg homeostasis in these bacteria has not been undertaken. Genes predicted to encode homologs of the high affinity Mg transporters CorA (Park et al. 1976; Hmiel et al. 1986), MgtE (Smith et al. 1995), and MgtB (Hmiel et al. 1989) can be found in *Brucella* genome sequences (Fig. 6.1). CorA and MgtE are gated channels, the interactions of Mg with the cytoplasmic portions of these proteins opens or closes the corresponding 'gates', and the influx of Mg into the cell is driven by the membrane potential. In contrast, MgtB is a P-type ATPase, and ATP hydrolysis and phosphorylation of MgtB provides the energy to support Mg^{2+} transport. Based on the characterization of these transporters in other bacteria (Papp-Wallace and Maguire 2008; Groisman et al. 2013), it seems likely that CorA and MgtE are the major constitutive Mg transporters in *Brucella*. MgtB, on the other hand, may only be utilized when Mg concentrations in the external environment are limited. In *Salmonella*, for instance, the two component regulator PhoPQ recognizes external Mg levels and activates *mgtB* expression in response to Mg limitation (Soncini et al. 1996). Notably, a *B. melitensis* mTn5 mutant with an insertion in *mgtB* has been reported to be attenuated in mice and cultured mammalian cells (Lestrate et al. 2000). But as will be described in the next section, the role of MgtB in virulence is presently unclear due to the possibility of a polar effect on a downstream gene in the *B. melitensis mgtB*::mTn5 mutant.

6.1.2 MgtC

The cytoplasmic membrane protein MgtC is an important virulence determinant in *Salmonella* (Blanc-Potard and Groisman 1997). This protein was originally thought to be involved in Mg transport, but recent studies have shown that it actually inhibits the activity of the F_1F_0 ATPase (Lee et al. 2013), which is thought to be important for maintaining the proper cellular balance of Mg and ATP during periods of Mg deprivation and preventing disruption of ribosome function (Pontes et al. 2015a). MgtC also has an indirect effect on c-dGMP signaling that prevents exopolysaccharide production by *S. enterica* serovar Typhimurium when this bacterium is residing host macrophages (Pontes et al. 2015b). The *Salmonella* genes encoding MgtC and the Mg transporter MgtB reside in an operon (Fig. 6.2), and PhoPQ-mediated induction of *mgtC* and *mgtB* in response to Mg limitation is considered to be an important adaptation of *Salmonella* strains to the intracellular environment in their mammalian hosts (Groisman et al. 2013).

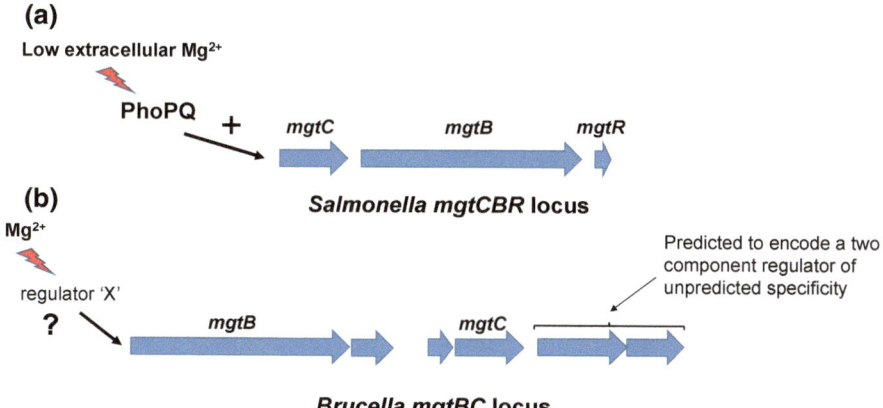

Fig. 6.2 **a** Genetic organization and PhoPQ-mediated regulation of the *Salmonella mgtCBR* locus and **b** genetic organization of the *Brucella mgtBC* locus

Brucella strains produce an MgtC homolog, and this protein is required for the wild-type growth of *B. suis* 1330 in a Mg-deficient medium and the virulence of this strain in cultured mammalian cells (Lavigne et al. 2005). But whether or not the *Brucella* MgtC performs the same function as its *Salmonella* counterpart (e.g. inhibition of ATPase activity) remains to be determined experimentally. It is interesting, however, that the genetic organization of the *Brucella mgtB* and *mgtC* genes (Fig. 6.2) suggests that these genes may be transcribed as an operon like their counterparts are in *Salmonella*. mTn5 insertions typically produce polar mutations, which raises the possibility that the attenuation reported for the *B. melitensis mgtB::*mTn5 mutant may have been the result of a polar effect on the downstream *mgtC* gene. Thus, it will be important to employ a non-polar mutagenesis strategy to define the role of the *Brucella mgtB* in virulence.

The genetic organization of the *Brucella mgtB* and *mgtC* also suggests that these genes might be co-transcribed under the control of a transcriptional regulator that recognizes extracellular Mg levels in the same fashion that PhoPQ regulates the homologous genes in *Salmonella* (Groisman et al. 2013). The *Brucella* two component regulator FeuPQ was originally thought to be a PhoPQ homolog (Dorrell et al. 1998), but phenotypic evaluation of a *B. suis feuP* mutant did not provide evidence to support this proposition. But it is notable that the two genes immediately downstream of the *Brucella mgtC* (Fig. 6.2) are predicted to encode a two component regulator of unpredicted function.

6.2   Copper

The redox activity of copper (Cu) allows this metal to serve as an essential co-factor for cytochrome oxidases and denitrification enzymes in bacterial respiratory chains (Argüello et al. 2013). Cu, Zn-cofactored superoxide dismutases are also important antioxidants in many bacteria (De Groote et al. 1997; Gort et al. 1999; Piddington et al. 2001). But in contrast to Mg, Cu sits at the top of the Irving-Williams scale of metal reactivity (Foster et al. 2014), and the highly reactive nature of 'unbound' Cu can damage Fe-S clusters (Macomber and Imlay 2009) and generate ROS via Fenton chemistry (Ladomersky and Petris 2015). Consequently, most bacteria are not thought to actively import Cu into their cytoplasm (Tottey et al. 2005), and the limited number of bacterial Cu-containing proteins they contain are usually either integrated into the cytoplasmic membrane with their Cu-centers oriented toward the exterior, or located in the periplasmic compartment of Gram-negative bacteria (Argüello et al. 2013).

In addition to this protective compartmentalization of Cu-containing proteins, bacteria also rely upon the concerted activities of Cu chaperones like CopZ (Cobine et al. 1999) and Cu efflux pumps such as CopA (Rensing et al. 2000) to actively remove any unbound Cu from the cytoplasm (González-Guerrero and Argüello 2008), and the genes that encode these proteins are controlled by highly sensitive Cu-responsive regulators. The transcriptional regulator CueR that activates the expression of *copA* and *copZ* in *E. coli*, for instance, has an affinity for Cu so low, e.g. 10^{-21} M, that is has been proposed that this efflux system ensures that there is essentially no 'free' Cu in the cytoplasm of this bacterium (Changela et al. 2003). Gram-negative bacteria also have dedicated Cu exporters such as FixI (Preisig et al. 1996) and GolT (Osman et al. 2013) and periplasmic Cu chaperones such as PcuC (Serventi et al. 2012), SenC (Lohmeyer et al. 2012) and CueP (Osman et al. 2013) to ensure that Cu exported to the periplasm is directly incorporated into Cu-containing proteins such as cytochrome oxidases or SodC. Multicopper oxidases such as CueO also limit the toxicity of any 'free' Cu in the periplasm by converting it from Cu^+ to Cu^{2+} (Grass and Rensing 2001; Outten et al. 2001).

Although Cu has been reported to be toxic for *Brucella* strains (McCullough et al. 1947; Bleichert et al. 2014), Cu-containing proteins perform important functions in these bacteria. The Cu,Zn co-factored superoxide dismutase SodC, for instance, protects the brucellae from the oxidative burst of host macrophages (Gee et al. 2005). The CcoN subunit of cbb_3-type cytochrome *c* oxidase, the NirK subunit of nitrite reductase and the NosZ subunit of nitrous oxide reductase are also integral components of electron transport chains (Viebrock and Zumft 1988; Glockner et al. 1993; Thöny-Meyer et al. 1994) required for the respiratory metabolism of *Brucella* strains as well as their virulence in mice (Haine et al. 2006; Jiménez de Bagüés et al. 2007). Like most other bacteria, the brucellae do not appear to have genes dedicated to Cu import, but they do have homologs of several genes that have been linked to Cu resistance, e.g. *cueR*, *copA*, *copZ*, *cueO*, *fixI*, *senC* and *pcuC* (Fig. 6.3). But only in the case of the *cueO* homolog has an investigation of the functionality of these genes been reported. Wu et al. (2015) purified the CueO homolog BmcO from *B. melitensis* 16 M, showed that this protein has multicopper oxidase activity in vitro, and found that a *bmcO* mutant exhibits increased sensitivity to $CuCl_2$. Interestingly, these authors also found that BmcO has ferroxidase activity (e.g. $Fe^{2+} \rightarrow Fe^{3+}$) in vitro, which is potentially relevant because it infers that in addition to its role in Cu detoxification, the *Brucella* CueO homolog might also play a role in Fe transport like its counterpart does in *Pseudomonas aeruginosa* (Huston et al. 2002).

It has recently become appreciated that mammalian hosts employ Cu intoxication as a defense against microbial infections (Samanovic et al. 2012). Activation of host macrophages by interferon-γ (INF-γ), for instance, causes an increased influx of Cu into these phagocytes and increased transport of Cu into pathogen-containing phagosomes (White et al. 2009) (Fig. 6.4). Accordingly, copper resistance genes have been shown to be important virulence determinants for *Mycobacterium tuberculosis* (Ward et al. 2010; Wolschendorf et al. 2011), *Strepotococcus pneumoniae* (Shafeeq et al. 2011; Johnson et al. 2015), and *Salmonella enterica* serovar Typhimurium (Achard et al. 2010) in rodent models. But a *Brucella melitensis bmcO* is not attenuated in cultured murine macrophages (Wu et al. 2015), nor is a *B. abortus bmcO* mutant attenuated in C57BL/6 mice (data not shown). At first glance, these observations suggest that Cu intoxication does play a role in host defense against *Brucella* infections. But it will be important to examine the virulence properties of *Brucella* mutants lacking Cu detoxification proteins in addition to

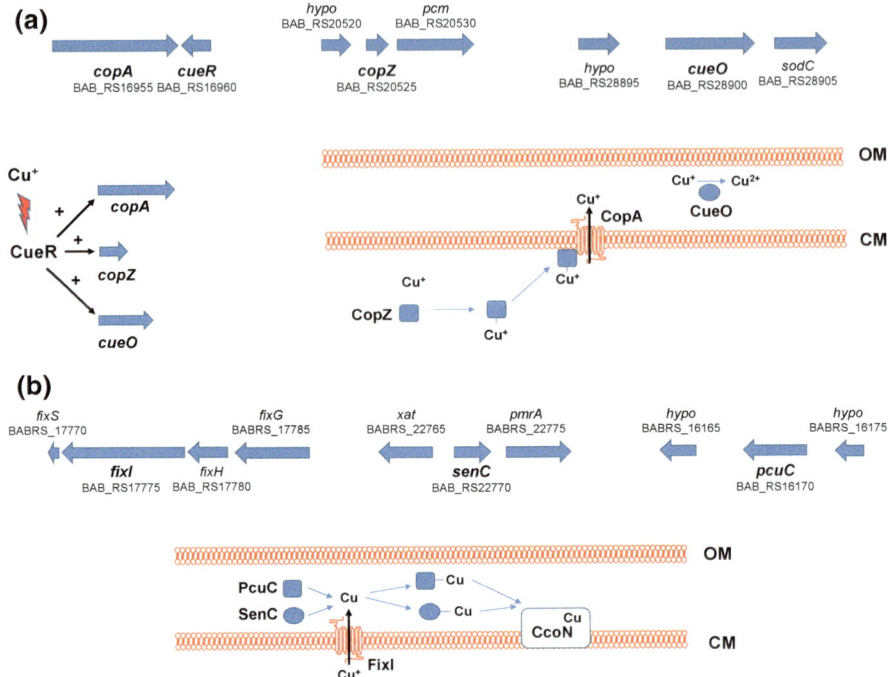

Fig. 6.3 Proposed roles of the *Brucella* CueR, CopZ, CopA and CueO (**a**) and FixI, SenC and PcuC (**b**) in preventing Cu toxicity. The corresponding gene designations are those used in the *B. abortus* 2308 genome sequence. OM—outer membrane; CM—cytoplasmic membrane; CcoN—subunit N of *cbb₃*-type cytochrome *c* oxidase

BmcO before any firm conclusions can be made in this regard. It is conceivable, however, that the intracellular trafficking of the *Brucella*-containing vacuoles in host macrophages (Celli 2015) prevents their interaction with the host cell Cu transporter ATP7A (White et al. 2009), providing a mechanism to avoid Cu intoxication.

6.3   Cobalt

Cobalt is an essential micronutrient for many bacteria including *Brucella* because it is an integral component of cobalamin (vitamin B_{12}) (Warren et al. 2002), and several important bacterial enzymes such as the methionine synthase MetH (Childs and Smith 1969) and the ribonucleotide reductase NrdJ (Cowles et al. 1969; Taga and Walker 2010) require cobalamin for their activity. *Brucella* strains possess an intact cobalamin biosynthesis pathway (Lawrence et al. 2008; Lundqvist et al. 2009), and *B. melitensis* and *B. suis* mutants with disruptions in this pathway are

Fig. 6.4 Cu intoxication of bacterial pathogens by IFN-γ-activated macrophages. CTR1 and ATP7A are transmembrane Cu transporters and ATOX1 is a cytoplasmic Cu chaperone

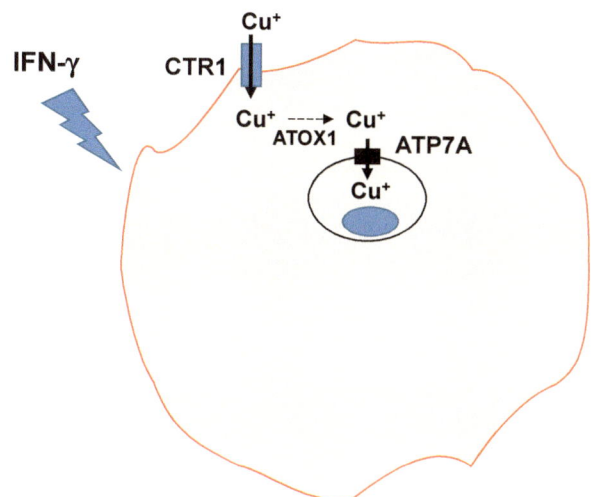

attenuated in mice and cultured mammalian cells (Köhler et al. 2002; Delrue et al. 2004). A *B. melitensis metH* mutant is also attenuated in these infection models (Lestrate et al. 2000), and moreover, TnSeq analysis suggests that *nrdJ* is likely an essential gene in *Brucella* (X. De Bolle, personal communication).

Two genes predicted to encode a CbtAB-type Co transporter reside between the cobalamin biosynthetic genes *cobQ* and *cobU* in *Brucella* (Fig. 6.5). This location, coupled with the fact that the *Brucella cbtA* and B genes reside downstream of a cobalamin-responsive riboswitch (Rodionov et al. 2003), supports the contention that they encode a high affinity Co transporter, but this function has not been confirmed experimentally. Interestingly, clusters of genes similar to those encoding the *Rhodobacter capsulatus* Co transporter CbiMNOQ (Siche et al. 2010) and the *Sinorhizobium meliloti* Co transporter CbtJKL (Cheng et al. 2011) can also be found in *Brucella* genomes, but experimental evidence indicates that these genes encode Ni (Sangari et al. 2010) and heme (Ojeda 2012) transporters, respectively, rather than Co transporters.

Like the case with most other metals, excess intracellular Co can be toxic to bacterial cells. The reactivity of Co falls between that of Ni and Fe on the Irving-Williams scale (Foster et al. 2014), and Co toxicity is thought to be largely related to its capacity to displace Fe and interfere with the proper assembly of Fe-S clusters in cellular proteins (Barras and Fontecave 2011). Co sensitivity has been reported for *Brucella* strains (Altenbern et al. 1959; Anderson et al. 2009), but there have been no reports describing experimental approaches designed to determine how these bacteria resist Co toxicity. The cobalamin riboswitch upstream of the *Brucella cbtA* and *B* genes conceivably provides these bacteria with a mechanism for preventing high affinity Co transport once the cell has acquired sufficient levels of this metal to support its need for cobalamin (Fig. 6.5). The CobS and CobT subunits of the cobaltochelatase complex in *Brucella* have also been proposed to

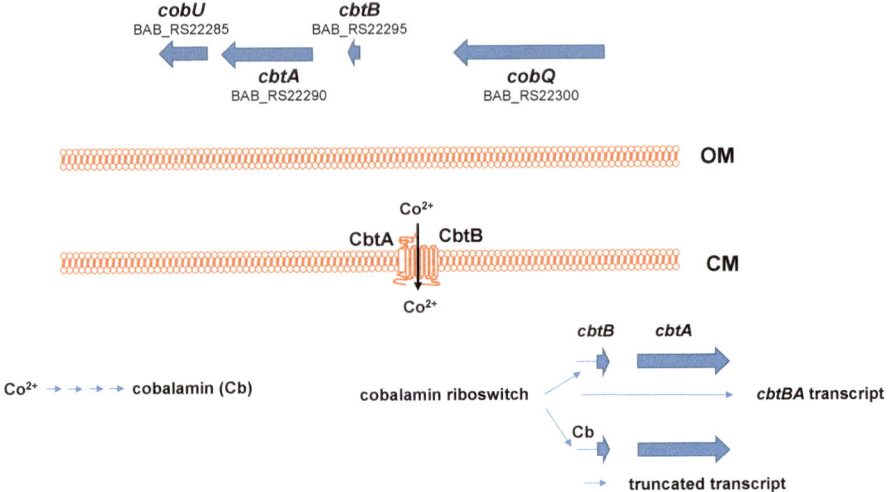

Fig. 6.5 Proposed role of the *Brucella* CbtAB in Co^{2+} import and repression of the corresponding genes by a cobalamin riboswitch. The corresponding gene designations are those used in the *B. abortus* 2308 genome sequence. Cb—cobalamin; OM—outer membrane; CM—cytoplasmic membrane

serve as a Co chaperone during cobalamin biosynthesis (Lundqvist et al. 2009). Other genes that may be linked to Co resistance in *Brucella* are putative *rcnA* and *dmeF* homologs. RcnA is a well-characterized Co and Ni exporter in *E. coli* (Rodrigue et al. 2005), and this exporter works in conjunction with the periplasmic substrate-binding protein RcnB to facilitate export of these cations (Blériot et al. 2011). The E. coli *rcnA* and *B* genes reside next to *rcnR*, which encodes a Co- and Ni-responsive regulator that controls *rcnA* and *B* expression (Iwig et al. 2006) (Fig. 6.6a). Homologs of *rcnB* and *rcnR* are not readily apparent in *Brucella* genomes, but it is notable that the *Brucella rcnA* resides next to genes annotated as coding for a periplasmic substrate binding protein of undefined specificity and a LysR-type transcriptional regulator (Fig. 6.6b). It is possible that the products of these two genes perform the same functions as RcnB and RcnR. The gene designated BAB_RS18410 in the *B. abortus* 2308 genome is also annotated as encoding a homolog of DmeF, a metal exporter that has been linked to Co and Ni resistance in the related α-proteobacteria *Agrobacterium tumefaciens* (Dokpikul et al. 2016) and *Rhizobium leguminosarum* (Rubio-Sanz et al. 2013). DmeF is a Co exporter and DmeR is a transcriptional regulator that induces *dmeF* expression in response to exposure to this metal. Considering that Co is such an important micronutrient for *Brucella* strains, it will be imperative to determine whether or not the proposed *rcnA* and *dmeF* do indeed contribute to Co homeostasis, and identify any other genes that contribute to this process.

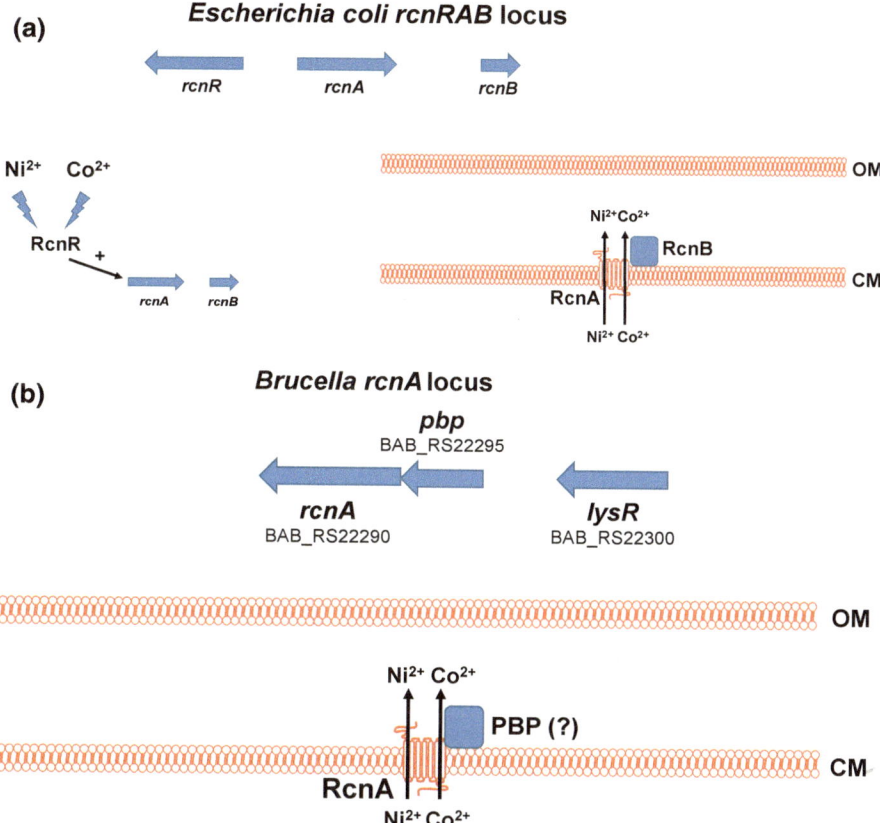

Fig. 6.6 Role of RcnA, RcnB and RcnR in Co^{2+} and Ni^{2+} export in *Escherichia coli* (**a**) and proposed role of RcnA in Co^{2+} and Ni^{2+} export in *Brucella* (**b**). The *Brucella* gene designations are those used in the (**b**). *abortus* 2308 genome sequence. OM—outer membrane; CM—cytoplasmic membrane

6.4 Cobalamin Transport

In addition to transporting Co and synthesizing cobalamin, there is also genomic evidence suggesting that *Brucella* strains can transport pre-formed cobalamin from the surrounding environment. Specifically, the genes designated BAB_RS22455-22470 in the *B. abortus* 2308 genome are annotated as encoding the TonB-dependent cobalamin transporter BtuBFCD (Heller et al. 1985; DeVeaux and Kadner 1985; Cadieux et al. 2002). This proposition is also supported by fact that these genes appear to be preceded by a cobalamin riboswitch (Rodionov et al. 2003). However, as noted in Chap. 2 of this text, these genes have also been proposed to be involved in the siderophore transport (Roop 2012; Roop et al. 2004, 2012). It will be obviously be important to clearly define the specificity of the transporter encoded by these genes in future studies.

6.5 Conclusions

Mg, Cu and Co play important roles in basic biology and virulence of *Brucella* strains, but to date little experimental work has been done to determine how these metals are transported, stored and detoxified in these bacteria. The potential role of Cu intoxication as a host defense against *Brucella* strains has also not been thoroughly examined. Consequently, these areas of research offer the opportunity to not only better understand how these bacteria control their metal homeostasis, but they also provide the opportunity to gain better insight into how *Brucella* strains produce disease in their mammalian hosts.

References

Achard MES, Tree JJ, Holden JA, Simpfendorfer KR, Wijburg OLC, Strugnell RA, Schembri MA, Sweet MJ, Jennings MP, McEwan AG (2010) The multi-copper-ion oxidase CueO of *Salmonella enterica* serovar Typhimurium is required for systemic virulence. Infect Immun 78:2312–2319

Altenbern RA, Williams DR, Ginoza HS (1959) Effect of cobalt on population changes in *Brucella abortus*. J Bacteriol 77:509

Anderson ES, Paulley JT, Gaines JM, Valderas MW, Martin DW, Menscher E, Brown TD, Burns CS, Roop RM II (2009) The manganese transporter MntH is a critical virulence determinant for *Brucella abortus* 2308 in experimentally infected mice. Infect Immun 77:3466–3474

Andreini C, Bertini I, Cavallaro G, Holliday GL, Thornton JM (2008) Metal ions in biological catalysis: from enzyme databases to general principles. J Biol Inorg Chem 13:1205–1218

Argüello JM, Raimunda D, Padilla-Benavides T (2013) Mechanisms of copper homeostasis in bacteria. Front Cell Infect Micrbiol 3:e73

Barras F, Fontecave M (2011) Cobalt stress in *Escherichia coli* and *Salmonella enterica*: molecular bases for toxicity and resistance. Metallomics 3:1130–1134

Blanc-Potard AB, Groisman EA (1997) The *Salmonella selC* locus contains a pathogenicity island mediating intramacrophage survival. EMBO J 16:5376–5385

Bleichert P, Santo CE, Hanczaruk M, Meyer H, Grass G (2014) Inactivation of bacterial and viral biothreat agents on metallic copper surfaces. Biometals 27:1179–1189

Blériot C, Effantin G, Lagarde F, Mandrand-Berthelot MA, Rodrigue A (2011) RcnB is a perisplamic protein essential for maintaining intracellular Ni and Co concentrations in *Escherichia coli*. J Bacteriol 193:3785–3793

Cadieux N, Bradbeer C, Reeger-Schneider E, Koster W, Mohanty AK, Wiener MC, Kadner RJ (2002) Identification of the periplasmic cobalamin-binding protein BtuF of *Escherichia coli*. J Bacteriol 184:706–717

Celli J (2015) The changing nature of the *Brucella*-containing vacuole. Cell Microbiol 17:951–958

Changela A, Chen K, Xue Y, Holschen J, Outten CE, O'Halloran TV, Mondragón A (2003) Molecular basis of metal-ion selectivity and zeptomolar sensitivity by CueR. Science 301:1383–1387

Cheng J, Poduska B, Morton RA, Finan TM (2011) An ABC-type cobalt transport system is essential for growth of *Sinorhizobium meliloti* at trace metal concentrations. J Bacteriol 193:4405–4416

Childs JD, Smith DA (1969) New methionine structural gene in *Salmonella typhimurium*. J Bacteriol 100:377–381

Cobine P, Wickramasinghe WA, Harrison MD, Weber T, Solioz M, Dameron CT (1999) The *Enterococcus hirae* copper chaperone CopZ delivers copper (I) to the CopY repressor. FEBS Lett 445:27–30

Cowles JR, Evans HJ, Russell SA (1969) B_{12} coenzyme-dependent ribonucleotide reductase in *Rhizobium* species and the effects of cobalt deficiency on the activity of the enzyme. J Bacteriol 97:1460–1465

De Groote MA, Oschner UA, Shiloh MU, Nathan C, McCord JM, Dinauer MC, Libby SJ, Vazquez-Torres A, Xu Y, Fang FC (1997) Periplasmic superoxide dismutase protects *Salmonella* from products of phagocyte NADPH-oxidase and nitric oxide synthase. Proc Natl Acad Sci USA 94:13997–14001

Delrue RM, Lestrate P, Tibor A, Letesson JJ, De Bolle X (2004) *Brucella* pathogenesis, genes identified from random large-scale screens. FEMS Microbiol Lett 231:1–12

DeVeaux LC, Kadner RJ (1985) Transport of vitamin B_{12} in *Escherichia coli*: cloning of the *btuCD* region. J Bacteriol 162:888–896

Dokpikul T, Chaoprasid P, Saninjuk K, Sirirakphaisarn S, Johnrod J, Nookabkaew S, Sukchawalit R, Mongkolsuk S (2016) Regulation of the cobalt/nickel efflux operon *dmeRF* in *Agrobacterium tumefaciens* and a link between the iron-sensing regulator RirA and cobalt/nickel resistance. Appl Environ Microbiol 82:4732–4742

Dorrell N, Spencer S, Foulongne V, Guigue-Talet P, O'Callaghan D, Wren BW (1998) Identification, cloning and initial characterization of FeuPQ in *Brucella suis*: a new sub-family of two-component regulatory systems. FEMS Microbiol Lett 162:143–150

Evenson MA, Gerhardt P (1955) Nutrition of brucellae: utilization of iron, magnesium and manganese for growth. Proc Soc Exp Biol Med 89:678–680

Foster AW, Osman D, Robinson NJ (2014) Metal preferences and metallation. J Biol Chem 289:28095–28103

Gee JM, Valderas MW, Kovach ME, Grippe VK, Robertson GT, Ng WL, Richardson JM, Winkler ME, Roop RM II (2005) The *Brucella* Cu, Zn superoxide dismutase is required for optimal resistance to oxidative killing by murine macrophages and wild-type virulence in experimentally infected mice. Infect Immun 73:2873–2880

Glockner AB, Jüngst A, Zumft WG (1993) Copper-containing nitrite reductase from *Pseudomonas aureofaciens* is functional in a mutationally cytochrome cd_1-free background (NirS-) of *Pseudomonas stutzeri*. Arch Microbiol 160:18–26

González-Guerrero M, Argüello JM (2008) Mechanism of Cu^+-transporting ATPases: soluble Cu^+ chaperones directly transfer Cu^+ to transmembrane transport sites. Proc Natl Acad Sci USA 105:5992–5997

Gort AS, Ferber DM, Imlay JA (1999) The regulation and role of the periplasmic copper, zinc superoxide dismutase of *Escherichia coli*. Mol Microbiol 32:179–191

Grass G, Rensing C (2001) CueO is a multi-copper oxidase that confers copper tolerance in *Escherichia coli*. Biochem Biophys Res Comm 286:902–908

Groisman EA, Hollands J, Kriner MA, Lee EJ, Park SY, Pontes MH (2013) Bacterial Mg^{2+} homeostasis, transport and virulence. Annu Rev Genet 47:625–646

Haine V, Dozot M, Dornand J, Letesson JJ, De Bolle X (2006) NnrA is required for full virulence and regulates several *Brucella melitensis* denitrification genes. J Bacteriol 188:1615–1619

Heller K, Mann BJ, Kadner RJ (1985) Cloning and expression of the gene for the vitamin B_{12} receptor protein in the outer membrane of *Escherichia coli*. J Bacteriol 161:896–903

92   R.M. Roop II et al.

Hmiel SP, Snavely MD, Miller CG, Maguire ME (1986) Magnesium transport in *Salmonella typhimurium*: characterization of magnesium influx and cloning of a transport gene. J Bacteriol 168:1444–1450

Hmiel SP, Snavely MD, Florer JB, Maguire ME, Miller CG (1989) Magnesium transport in *Salmonella typhimurium*: genetic characterization and cloning of three magnesium transport loci. J Bacteriol 171:4742–4751

Huston WM, Jennings MP, McEwan AG (2002) The multicopper oxidase of *Pseudomonas aeruginosa* is a ferroxidase with a central role in iron acquisition. Mol Microbiol 45:1741–1750

Iwig JS, Rowe JL, Chivers PT (2006) Nickel homeostasis in *Escherichia coli*—the rcnR-rcnA efflux pathway and its linkage to NikR function. Mol Microbiol 62:252–262

Jiménez de Bagüés MP, Loisel-Meyer S, Liautard JP, Jubier-Maurin V (2007) Different roles of the two high-oxygen-affinity terminal oxidases of *Brucella suis*: cytochrome c oxidase, but not ubiquinol oxidase, is required for persistence in mice. Infect Immun 75:531–535

Johnson MDL, Kehl-Fie TE, Klein R, Kelly J, Burnham C, Mann B, Rosch JW (2015) Role of copper efflux in pneumococcal pathogenesis and resistance to macrophage-mediated immune clearance. Infect Immun 83:1684–1694

Köhler S, Foulongne V, Ouahrani-Bettache S, Bourg G, Teyssier J, Ramuz R, Liautard JP (2002) The analysis of the intramacrophagic virulome of *Brucella suis* deciphers the environment encountered by the pathogen inside the macrophage host cell. Proc Natl Acad Sci USA 99:15711–15716

Ladomersky E, Petris MJ (2015) Copper tolerance and virulence in bacteria. Metallomics 7:957–964

Lavigne JP, O'Callaghan D, Blanc-Potard AB (2005) Requirement of MgtC for *Brucella suis* intramacrophagic growth: a potential mechanism shared by *Salmonella enterica* and *Mycobacterium tuberculosis* for adaptation to a low-Mg^{2+} environment. Infect Immun 73:3160–3163

Lawrence AD, Deery E, McLean KJ, Munro AW, Pickersgill RW, Rigby SEJ, Warren MJ (2008) Identification, characterization, and structure/function analysis of a corrin reductase involved in adenosylcobalamin biosynthesis. J Biol Chem 283:10813–10821

Lee EJ, Pontes MH, Groisman EA (2013) A bacterial virulence protein promotes pathogenicity by inhibiting the bacterium's own F_1F_0 ATP synthase. Cell 154:146–156

Lestrate P, Delrue RM, Danese I, Didembourg C, Taminiau B, Mertens P, De Bolle X, Tibor A, Tang CM, Letesson JJ (2000) Identification and characterization of *in vivo* attenuated mutants of *Brucella melitensis*. Mol Microbiol 38:543–551

Lohmeyer E, Schroder S, Pawlik G, Trasnea PI, Peters A, Daldal F, Koch HG (2012) The ScoI homologue SenC is a copper binding protein that interacts directly with the cbb₃-type cytochrome oxidase in *Rhodobacter capsulatus*. Biochim Biophys Acta 1817:2005–2015

Lundqvist J, Elmlund D, Heldt D, Deery E, Söderberg CAG, Hansson M, Warren M, Al-Karadaghi S (2009) The AAA^+ motor complex of subunits CobS and CobT of cobaltochelatase visualized by single particle electron microscopy. J Structural Biol 167:227–234

Macomber L, Imlay JA (2009) The iron-sulfur clusters of dehydratases are primary intracellular targets of copper toxicity. Proc Natl Acad Sci USA 106:8344–8349

McCullough WG, Mills RC, Herbst EJ, Roessler WG, Brewer CR (1947) Studies on the nutritional requirements of *Brucella suis*. J Bacteriol 53:5–15

Ojeda JF (2012) The bhuTUV and bhuO genes play vital roles in the ability of *Brucella abortus* to use heme as an iron source and are regulated in an iron-responsive manner by RirA and Irr. Doctoral dissertation, East Carolina University

Osman D, Patterson CJ, Bailey K, Fisher K, Robinson NJ, Rigby SEJ, Cavet JS (2013) The copper supply pathway to a *Salmonella* Cu, Zn-superoxide dismutase (SodCII) involves P1B-type ATPase copper efflux and periplasmic CueP. Mol Microbiol 87:466–477

Outten FW, Huffman DL, Hale JA, O'Halloran TV (2001) The independent *cue* and *cus* systems confer copper tolerance during aerobic and anaerobic growth in *Escherichia coli*. J Biol Chem 276:30670–30677

Papp-Wallace KM, Maguire ME (2008) Magnesium transport and magnesium homeostasis. EcoSal Plus 2013. doi:10.1128/ecosalplus.5.4.4.2

Park MH, Wong BB, Lusk JE (1976) Mutants in three genes affecting transport of magnesium in *Escherichia coli*: genetics and physiology. J Bacteriol 126:1096–1103

Piddington DL, Fang FC, Laessig T, Cooper AM, Orme IM, Buchmeier NA (2001) Cu, Zn superoxide dismutase of *Mycobacterium tuberculosis* contributes to survival in activated macrophages that are generating an oxidative burst. Infect Immun 69:4980–4987

Pontes MH, Sevostyanova A, Groisman EA (2015a) When too much ATP is bad for protein synthesis. J Mol Biol 427:2586–2594

Pontes MH, Lee EJ, Choi J, Groisman EA (2015b) *Salmonella* promotes virulence by repressing cellulose production. Proc Natl Acad Sci USA 112:5183–5188

Preisig O, Zufferey R, Hennecke H (1996) The *Bradyrhizobium japonicum fixGHIS* genes are required for the formation of high-affinity cbb_3-type cytochrome oxidase. Arch Microbiol 165:297–305

Rensing C, Fan B, Sharma R, Mitra B, Rosen BP (2000) CopA: an *Escherichia coli* Cu (I)-translocating P-type ATPase. Proc Natl Acad Sci USA 97:652–656

Rodionov DA, Vitreschak AG, Mironov AA, Gelfand MS (2003) Comparative genomics of the vitamin B_{12} metabolism and regulation in prokaryotes. J Biol Chem 278:41148–41159

Rodrigue A, Effantin G, Mandrand-Berthelot MA (2005) Identification of *rcnA* (*yohM*), a nickel and cobalt resistance gene in *Escherichia coli*. J Bacteriol 187:2912–2916

Roop RM II (2012) Metal acquisition and virulence in *Brucella*. Anim Health Res Rev 13:10–20

Roop RM II, Bellaire BH, Anderson ES, Paulley JT (2004) Iron metabolism in *Brucella*. In: López-Goñi I, Moriyón I (eds) *Brucella*—molecular and cellular biology. Horizon Bioscience, Norfolk, UK, pp 243–262

Roop RM II, Anderson E, Ojeda J, Martinson D, Menscher E, Martin DW (2012) Metal acquisition by *Brucella* strains. In: López-Goñi I, O'Callaghan D (eds) *Brucella*—molecular microbiology and genomics. Caister Academic Press, Norfolk, UK, pp 179–199

Rubio-Sanz L, Prieto RI, Imperial J, Palacios JM, Brito B (2013) Functional and expression analysis of the metal-inducible *dmeRF* system from *Rhizobium leguminosarum* bv. viciae. Appl Environ Microbiol 79:6414–6422

Samanovic MI, Ding C, Thiele DJ, Darwin KH (2012) Copper in microbial pathogenesis: meddling with the metal. Cell Host Microbe 11:106–115

Sanders TH, Higuchi K, Brewer CR (1953) Studies on the nutrition of *Brucella melitensis*. J Bacteriol 66:294–299

Sangari FJ, Cayón AM, Seoane A, García-Lobo JM (2010) *Brucella abortus ure2* region contains an acid-activated urea transporter and a nickel transport system. BMC Microbiol 10:e107

Serventi F, Youard ZA, Murset V, Huwiler S, Bühler D, Richter M, Luchsinger R, Fischer HM, Brogioli R, Niederer M, Hennecke H (2012) Copper starvation-inducible protein for cytochrome oxidase biogenesis in *Bradyrhizobium japonicum*. J Biol Chem 287:38812–38823

Shafeeq S, Yesilkaya H, Kloosterman TG, Narayanan G, Wandel M, Andrew PW, Kuipers OP, Morrissey JA (2011) The *cop* operon is required for copper homeostasis and contributes to virulence in *Streptococcus pneumoniae*. Mol Microbiol 81:1255–1270

Siche S, Neubauer O, Hebbeln P, Eitinger T (2010) A bipartite S unit of an ECF-type cobalt transporter. Res Microbiol 161:824–829

Smith RL, Thompson LJ, Maguire ME (1995) Cloning and characterization of MgtE, a putative new class of Mg^{2+} transporter from *Bacillus firmus* OF4. J Bacteriol 177:1233–1238

Soncini FC, Véscovi EG, Solomon F, Groisman EA (1996) Molecular basis of the magnesium deprivation response in *Salmonella typhimurium*: identification of PhoP-regulated genes. J Bacteriol 178:5092–5099

Swem DL, Swen LR, Setterdahl A, Bauer CE (2005) Involvement of SenC in assembly of cytochrome c oxidase in *Rhodobacter capsulatus*. J Bacteriol 187:8081–8087

Taga ME, Walker GC (2010) *Sinorhizobium meliloti* requires a cobalamin-dependent ribonucleotide reductase for symbiosis with its plant host. Mol Plant Microbe Interact 23:1643–1654

Thöny-Meyer L, Beck C, Preisig O, Hennecke H (1994) The *ccoNOQP* gene cluster codes for a *cb*-type cytochrome oxidase that functions in aerobic respiration in *Rhodobacter capsulatus*. Mol Microbiol 14:705–716

Tottey S, Harvie DR, Robinson NJ (2005) Understanding how cells allocate metals using metal sensors and metallochaperones. Acc Chem Res 38:775–783

Viebrock A, Zumft WG (1988) Molecular cloning, heterologous expression, and primary structure of the structural gene for the copper enzyme nitrous oxide reductase from denitrifying *Pseudomonas stutzeri*. J Bacteriol 170:4658–4668

Waldron KJ, Robinson NJ (2009) How do bacterial cells ensure that metalloproteins get the correct metal? Nature Rev Microbiol 6:25–35

Ward SK, Abomoelak B, Hoye EA, Steinberg H, Talaat AM (2010) CtpV: a putative copper exporter required for full virulence of *Mycobacterium tuberculosis*. Mol Microbiol 77:1096–1110

Warren MJ, Raux E, Schubert HL, Escalante-Semerena JC (2002) The biosynthesis of adenosylcobalamin (vitamin B_{12}). Nat Prod Rep 19:390–412

White C, Lee J, Kambe T, Fritsche K, Petris MJ (2009) A role for the ATP7A copper-transporting ATPase in macrophage bactericidal activity. J Biol Chem 284:33949–339956

Wolschendorf F, Ackart D, Shrestha TB, Hascall-Dove L, Nolan S, Lamichhane G, Wang Y, Bossmann SH, Basaraba RJ, Niederweis M (2011) Copper resistance is essential for virulence of *Mycobacterium tuberculosis*. Proc Natl Acad Sci USA 108:1621–1626

Wu T, Wang S, Wang Z, Peng X, Lu Y, Wu Q (2015) A multicopper oxidase contributes to the copper tolerance of *Brucella melitensis* 16M. FEMS Microbiol Lett 362:1–7